Engineering coatings
— design and application

Engineering coatings
— design and application

Stan Grainger

ABINGTON PUBLISHING
Woodhead Publishing Ltd in association with The Welding Institute
Cambridge   England

Published by Abington Publishing,
Woodhead Publishing Ltd, Abington Hall,
Abington, Cambridge CB1 6AH, England

First published 1989

© Woodhead Publishing Ltd

Conditions of sale
All rights reserved. No part of this publication may be reproduced or transmitted in any form or by any means, electronic or mechanical, including photocopy, recording, or any information storage and retrieval system, without permission in writing from the publisher.

British Library Cataloguing in Publication Data

A CIP catalogue record for this book is available from the British Library.

ISBN 1 85573 000 6

Designed, typeset and printed by Crampton & Sons Ltd, Sawston, Cambridge CB2 4BQ, England.

This publication has been sponsored by:
Coated Electrodes UK Ltd;
Höganäs Specialty Powder;
Metco Ltd;
Mixalloy Ltd;
Provacuum Ltd;
Soudometal.

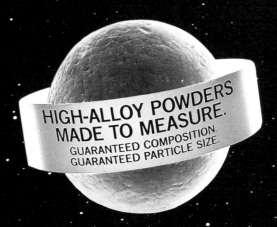

Contents

Foreword   11

Introduction   13
Surfacing materials and processes
Materials
  Processes
  Coating thickness
Material/process selection
Using the handbook

1   Mechanisms of wear   19
Basic mechanisms of wear
  Abrasive wear
  Adhesive wear
  Contact fatigue
Contributory processes and effects
  Fretting
  Erosion
  Corrosion
  Elevated temperature
Testing for wear resistance
Practical diagnosis of wear
  Abrasion
  Adhesion
  Contact fatigue
  Fretting
  Erosion and cavitation
  Corrosion
  Elevated temperature

2   Weld surfacing materials and processes   33
Characteristics of weld surfacing
  Coating thickness
  Adhesion to substrate
  Range of processes
  Resurfacing
Materials for weld surfacing and their selection
Welding processes for surfacing
  Process principles and their characteristics
  Power supplies for electric welding processes
Metallurgical and related effects of weld surfacing on substrate and coating
  The temperature cycle
  Practical recommendations
  Influence of substrate defects
  Dilution
  Coating properties and structure
  Buffer layers
Designing for weld surfacing
  Deposit thickness
  Blank preparation and deposit design
  Weld patterns
  Machining allowances
  Coating area
  Physical and mechanical properties of coatings
Practical considerations
  Manual versus mechanised surfacing
  Accessibility
  Weld defects
  Treatment of defects
Explosive weld cladding and friction surfacing

3 Thermal spraying processes and materials — 77
Thermal spray materials
  Materials with good bonding characteristics
  Ferrous alloys
  Nickel base alloys
  Cobalt base alloys
  Tungsten carbide and chromium carbide
  Ceramics
  Miscellaneous coating materials
  Cermets and graded coatings
Characteristics of thermal spray processes
  Oxyfuel gas wire spraying
  Oxyfuel gas powder spraying
  Electric arc wire spraying
  Plasma arc spraying in air (APS)
  Vacuum plasma spraying
  High velocity combustion spraying
Design for thermally sprayed coatings
  Component shape
  Surface profile details
  Deposit thickness
Coating production
  Surface preparation
  Bond coats
  Masking
  Sealing
  Spraying
Spray fused coatings
  Substrate materials
  Coating alloys
  Component design for spray fuse coatings
  Surface preparation before spraying
  Preheating
  Practical technique
Conclusions

4 Electrodeposited coatings — 101
Basic principles
  Vat plating
  Brush or selective plating
  Electroless or autocatalytic deposition
  Hard anodising
  Coating materials
Selection
Design
  Component design
  Process design for vat plating
Substrate materials
Surface preparation
Finishing
Specifications, inspection and quality assurance

5 Physical and chemical vapour deposition techniques — 119
The processes
  Surface preparation
Vacuum evaporation
  Evaporation sources
  Materials and applications
Gas scattering deposition
  Ion plating
  Activated reactive evaporation techniques (ARE)
Sputter coating
  Substrate biasing
  Gas ionisation
  Materials and applications
Ion implantation
  Materials and applications

Chemical vapour deposition
  Process characteristics
  Reaction types
  Plasma activated CVD
  Materials and applications
Comparison of processes
  Coating thickness
  Coating rate
  Cost
Design for physical and chemical vapour deposition
  Design requirements
  Surface condition
  Data to be supplied to the processor

6 Plastics coatings 139
Materials
Application processes
  Flame spraying
  Fluidised bed coating
  Electrostatic spraying
  Other techniques
Designing for plastics coatings
  Surface preparation
  Component design
Summary

7 Finishing of surface coatings applied by welding and thermal spraying 145
Fused and welded coatings
  The need for machining
  Turning
  Milling
  Drilling
  Grinding
  Honing
  Lapping
  Spark erosion
Thermally sprayed coatings
  Turning
  Grinding
  Other machining methods
  Plasma sprayed coatings

8 Quality assurance in surfacing 153
The quality plan
  Design
  Component/substrate requirements
  Consumables for welding and thermal spraying
  Surfacing practice
  Finishing
  Checklist of operations
Testing and inspection of coated components
  Weld deposited coatings
  Thermally sprayed coatings
Conclusions

9 Safe working in surfacing 163
Surfacing by welding and thermal spraying
  Compressed gases
  Fire
  Fumes and dust
  Noise
  Radiation
  Electric shock
  Surface preparation

Vacuum deposition
  PVD techniques
  CVD techniques
Electrodeposition
  Workflow
  Engineering
  Process materials
  Storage and preparation of working solutions
Bibliography
  Legislation and standards relevant to safe working practices in surfacing

10 Industrial applications of engineering coatings 175
The aircraft industry
Chemical and petroleum industries
Earthmoving, agricultural, quarrying and mining
Internal combustion (piston) engines
Food
Forging
Glass
Pulp and paper
Plastics and rubber
Power generation
Steelmaking
Textiles
Timber
Transport

11 Glossary of terms used in surfacing 189

List of sponsors 195

Index 197

Foreword

This engineering coatings handbook deals with materials and application processes used for improving the surface durability of engineering components in service. Surfacing technology has developed rapidly in recent years and, perhaps because of its specialised nature, its scope and the benefits it provides are not widely recognised.

The purpose of the handbook is to acquaint readers with the subject and to guide the choice of coatings and means of application to fit specific circumstances.

Production of this book was recommended by the Committee of the Surface Engineering Society which is affiliated to The Welding Institute. The source material includes technical papers provided by committee members and illustrations and information supplied by many industrial organisations and individual authors. All their contributions, too numerous to mention separately, are gratefully acknowledged, as is the help and advice provided by individual staff at The Welding Institute.

Stan Grainger
April 1989

Introduction

Progressive deterioration of metallic surfaces in use, commonly referred to as 'wear', ultimately leads to loss of plant operating efficiency and at worst a breakdown.

The cost to industry of wear to equipment is high and recognition of this fact lies behind the continuing development of the technology known as surface engineering which includes application of coatings to metal surfaces to improve their performance in specific working conditions. This is a subject of great importance to industry, which relies on long, trouble free operation of plant to obtain uniform product quality and lowest possible product cost.

This handbook, produced under the auspices of the Surface Engineering Society, brings together a wealth of information on coating materials and processes of application available, together with examples of how the technology is being used successfully in industry.

An important feature of engineering plant is that individual components must be designed with three objectives in mind:

— To present a series of surfaces to interact with other parts or with process media;
— To support these surfaces with adequate strength to withstand service stresses;
— To endow selected surfaces with resistance to wear.

The term 'wear' is used here to mean any of the destructive forms of attack which cause deterioration of metallic surfaces in use. These are described in detail in Chapter 1. Sometimes all surfaces of the part are exposed to wear but, more frequently, only a proportion of the surface is required to display high resistance to a specific form of attack.

The compositional requirements of the material necessary to provide adequate strength in a component are usually different from those which provide wear resistance, so a composite product consisting of a structural material with specially protected wearing surfaces naturally comes to mind.

Surfacing technology is used to cut the cost of component deterioration in service by providing:

— Acceptable service life/reduced downtime costs;
— Minimum first cost of the part consistent with the above, including use of the cheapest constructional materials and minimum use of special materials which provide wear resistance on selected surfaces;
— The opportunity, where possible, to repair the part after use by resurfacing.

Surfacing materials and processes

MATERIALS

Most metals, alloys, ceramics and some intermetallic compounds can be applied as coatings either individually or as mixtures, but their characteristics often limit the processes which can be used for their application. Refractory oxides, for example, cannot be applied by welding processes and carbides require a metal or alloy matrix.

The material/process relationship not only identifies where they can be used together but also determines properties that can be expected from the coating, such as density and adhesion to the substrate.

It is not possible to list all the surfacing consumables available for welding and thermal spraying; Table 0.1 summarises basic types. Each type may be available in more than one variation of composition, each of which has been developed to possess specific properties to optimise life in given circumstances. Tables 0.2 and 0.3 cover coatings applied by electro-deposition and by physical and chemical vapour deposition.

Table 0.1 Summary of consumables for surfacing by welding and thermal spraying

Group	Type	Weld deposition methods								Approximate deposit hardness, HV	Thermal spray methods								
		OA	MMA	MIG	FCAW	TIG	SAW		PTAW		Oxy-gas	Arc	Flasma crc	VPS/LPPS	'D' gun	Jet Kote	Diamond Jet	CDS	Spray fuse
1 Ferrous alloys	1 Carbon steels	✓	✓	✓		✓	✓			<250	✓	✓							
	2 Low alloy steels		✓	✓	✓		✓			250–650	✓	✓							
	3 Martensitic Cr steels		✓	✓	✓		✓			350–650			✓						
	4 High speed steels		✓	✓						600–700	✓								
	5 Austenitic stainless steels		✓	✓	✓	✓	✓			200/500*			✓			✓	✓		
	6 Austenitic Mn steels		✓	✓	✓					200/500*									
	7 Austenitic Cr-Mn steels		✓	✓	✓					200/600*									
	8 Austenitic irons		✓	✓	✓					300–600									
	9 Martensitic irons		✓	✓	✓					500–750									
	10 Hi Cr austenitic irons	✓	✓	✓	✓		✓		✓	500–750									
	11 Hi Cr martensitic irons	✓	✓	✓	✓		✓		✓	500–750									
	12 Hi Cr complex irons		✓	✓	✓					600–800									
	13 Fe-Cr-Co-Ni-Si									375–550	✓		✓						
2 Nickel alloys	1 Nickel		✓	✓						160	✓								
	2 Ni-Cu, Ni-Cu-In		✓	✓						130									
	3 Ni-Fe									200									
	4 Ni-Mo-Cr-W			✓	✓	✓				250/500*	✓	✓	✓		✓	✓	✓		
	5 Ni-Cr-Si-B (also with Cu+Mo)	✓				✓			✓	200–750	✓	(✓)	✓			✓			✓
	6 Ni-Mo-Fe									200–300	✓		✓						
	7 Ni-Al bond coat										✓	✓	✓			✓			
	8 Ni-Al-Cr bond coat										✓	✓	✓			✓			
	9 Ni-Cr 80/20											✓	✓		✓	✓			
	10 Ni-Cr-Fe						✓						✓			✓			
	11 M-Cr-Al-Y†			✓									✓	✓					
	12 Ni-Cr-Mo-Al-Ti												✓		✓				
3 Cobalt alloys	1 Co-Cr-W low alloy	✓	✓	✓		✓			✓	380–430			✓			✓			
	2 Co-Cr-W medium alloy	✓	✓	✓		✓			✓	480–550						✓			
	3 Co-Cr-W high alloy	✓	✓	✓		✓			✓	600–650						✓			
	4 Co-Cr-W-Ni	✓				✓			✓	390–450						✓			
	5 Co-Cr-Mo					✓			✓	300–350									
	6 Co-Cr-Mo-Si									300–700									
	7 Co-Cr-W-Ni-Si-B alloys								✓	400–700			✓						✓
4 Copper alloys	1 Brass									130	✓								
	2 Silicon bronze			✓						80–100									
	3 Aluminium bronze			✓						130–140	✓	✓					✓		
	4 Tin bronze			✓						40–110	✓								

Table 0.1 Contd

Group	Type	Weld deposition methods							Approximate deposit hardness, HV	Thermal spray methods								
		OA	MMA	MIG	FCAW	TIG	SAW	PTAW		Oxy-gas	Arc	Plasma arc	VPS/LPPS	'D' gun	Jet Kote	Diamond Jet	CDS	Spray fuse
5 Elemental metals	1 Aluminium									✓	✓	✓						
	2 Zinc									✓	✓							
	3 Tungsten											✓						
	4 Copper									✓	✓				✓			
	5 Molybdenum (also as bond coat)									✓		✓						
6 Composite materials	1 Ni-Cr-Si-B+Ni-Al or Mo									✓						✓		✓
	2 Ni-Cr-Si-B + carbides								600–750	✓								✓
	3 Co-Cr-W-Si-B + carbides			✓								✓						
	4 C steel + carbides	✓	✓															
	5 Ni + silicon carbide			✓														
	6 Tungsten carbide + cobalt								800–1100					✓	✓	✓	✓	
	7 Cr carbide + Ni-Cr								550–950			✓		✓	✓	✓	✓	
	8 Cermets									✓		✓		✓			✓	
	9 Cr-B††								750–1100	✓		✓						
7 Ceramics	1 Alumina									✓		✓						
	2 Zirconia									✓		✓						
	3 Chrome oxide									✓		✓						
	4 Alumina/titania									✓		✓		✓			✓	
	5 Titanium dioxide									✓		✓						
	6 Magnesium zirconate									✓		✓						
8 Plastics	1 Eva									(✓)°								
	2 Nylon									(✓)°								
	3 Polyester									(✓)°								
	4 Polypropylene									(✓)°								
	5 Low density polyethylene										(✓)°							

OA = oxyacetylene, MMA = manual metal arc, MIG = metal inert gas, FCAW = flux-cored arc welding, TIG = tungsten inert gas, SAW = submerged-arc welding, PTAW = plasma transferred arc welding, VPS/LPPS = vacuum plasma spraying/low pressure plasma spraying, D gun = detonation gun, Jet Kote = a trademark of Stoody Deloro Stellite Inc.

* Work hardened
† In the M-Cr-Al-Y alloy series, M covers Co, Ni and Fe singly and in combination
†† Supplied in paste form for preplacing and subsequent arc fusion to substrate
(✓) Ni-Cr-Si-B with Cu and Mo not suitable for these processes
° Other coating processes available for plastics

Table 0.2 Electrodeposited coatings

Type	Materials
Pure metals	Chromium, nickel, gold, silver, platinum, palladium, ruthenium, rhodium, lead, tin, iron
Alloys	Copper-tin, cobalt-tungsten, cobalt-molybdenum, tin-nickel
Composites	Cobalt-chromium carbide, nickel-silicon carbide

Table 0.3 Physical and chemical vapour coatings for wear resistance

Type	Materials
Elements	Carbon, silver, chromium, molybdenum, tungsten, nickel, titanium
Compounds	Carbides, nitrides, oxides, sulphides. Borides of: titanium, aluminium, silicon, tantalum, tungsten, chromium, molybdenum
Alloys	M-Cr-Al-Y alloys, cobalt-chromium

PROCESSES

Welding

Welding provides the highest bond strength between deposit and substrate, is capable of applying deposits of considerable thickness if required and, with a few exceptions, can be operated manually or be mechanised and programmed. A particular application of a given coating can often be achieved satisfactorily by more than one welding process and this flexibility of choice is a useful characteristic of the group. There is a wide range of materials available for welded coatings.

Thermal spraying

Thermal spraying offers two significant advantages over weld surfacing: the first is that non-weldable coating materials, such as ceramics, can be deposited as well as materials which are also weldable. The second is that all these materials can be deposited on substrate materials which are unsuited to welding because of their composition or because their thin section would distort excessively when welded.

Compared with welded coatings, thermally sprayed deposits exhibit some porosity and a lower strength of bond with the substrate. In general, deposit thickness is less than is possible by welding. Provided that these factors are taken into account at the design stage, thermal spraying can and does provide excellent and reliable service in applications as demanding as aircraft gas turbine engines.

Electrodeposition

A somewhat more limited range of coating materials is offered by electrodeposition but the low temperature of deposition provides advantages of low distortion, better access to internal surfaces and accurate control of deposit thickness.

Vapour deposition

Vapour deposition provides a limited range of coating material possibilities but can be used with materials difficult or impossible to apply by other techniques or to produce thin coatings of controlled thickness.

COATING THICKNESS

The thickness of coating material applied is normally consistent with the amount of wear that can be permitted before the component is no longer fit for use, and takes into account an allowance for machining or grinding the deposit to a specified dimension and finish. Table 0.4 gives a guide to ranges of thickness that are possible using the various processes; in practice a different range may result from the characteristics of a particular coating material.

Material/process selection

Faced with the wide range of possibilities indicated in the tables, selection of material and process may seem difficult, but is normally straightforward.

Table 0.4 Comparison of surfacing processes and deposits

	Vapour deposition	Electrodeposition	Thermal spraying	Spray fusing	Welding
Thickness, mm	0.001-0.2	0.02-0.5	0.1-1.0	0.5-1.5	1-20 or more
Component geometry	Versatile	Versatile	Access to internal surfaces controlled by size of torch/gun		
Component size	Limited by chamber size	Limited by plating bath	No limit	Limited by fusing facility	No limit
Substrate material	Almost limitless	Almost limitless	Almost limitless	Metals or alloys of higher melting point than coating	
Substrate temperature, °C	30-1000	100	200	1050	1400
Pretreatment	PVD — ion bombardment, CVD — various	Chemical cleaning and etching	Clean and roughen surface		Mechanical cleaning
Post-treatment	None/stress relief	None/stress relief	None	Substrate anneal/stress relief as required	
Coating porosity, %	Nil to small	Nil to small	1-15	Nil	Nil
Bond strength, MPa	High	100	20-140	High	High
Bond mechanism	Atomic, surface forces	Surface forces	Mechanical	Metallurgical	Metallurgical
Control of deposit thickness	Good	Good	Fairly good	Moderate	Manual — variable Mechanised — good
Distortion of substrate	Low	Low	Low	Moderate	Can be high, depending on substrate geometry

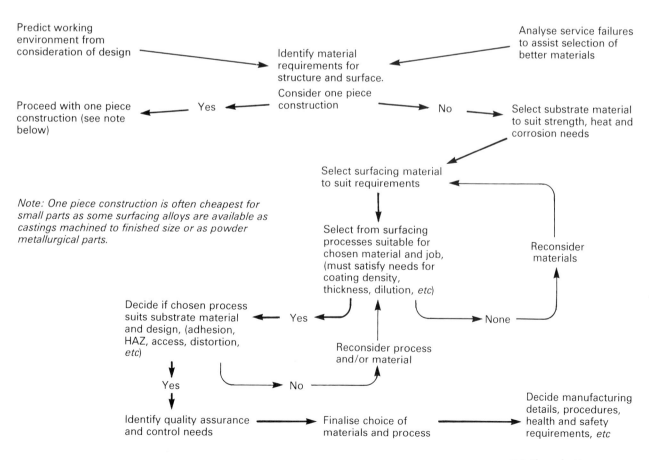

0.1 *The selection process.*

Often there are constraints placed on the choice because of availability. In many cases there is a precedent, but when considering a new problem it helps to follow a check list of the type shown in Fig.0.1.

The sequence of decisions to be made covers several fundamental points. The first is the need to be clear about service conditions, based on experience or plant design data. This is the key to material selection.

The second decision is choice of application process for that material. This will involve questions of compatibility with the coating material; reference to the tables in this chapter will show that not all the materials can be applied by all processes.

A further question of compatibility arises between both material and process with the substrate. All these issues are covered in subsequent chapters.

Using the handbook

The chapter devoted to mechanisms of wear is intended to assist in analysis of previous component failures and assessment of how parts of newly designed equipment are likely to suffer in service. In conjunction with later chapters, which give information on the ability of various coating materials to resist different forms of wear, a search can be made for a suitable coating material.

Four chapters deal with the main groups of coating processes outlined earlier. Those dealing with weld surfacing and thermal spraying are dealt with in some detail, recognising the widespread availability and use of these processes in industrial maintenance and production shops. Each chapter also describes consumable materials available for deposition by the processes. Application of plastics materials by thermal spraying is discussed in a separate chapter.

Wear resistant materials are more difficult to machine than steels and other alloys used for component construction, but precision finishing of most coatings is possible using appropriate techniques described in Chapter 7. Successful service of engineering coatings depends, as with other materials, on quality assurance and control. Special features of coating materials and processes which are important in this regard are dealt with in Chapter 8. The practice of surface coating uses combinations of materials and processes which differ from those met frequently in industry. It has been considered important therefore to include a section on safe working practice (Chapter 9).

Much can be learned from successful uses of surface coatings in industry. The chapter dealing with this illustrates the wide field of use to which the technology is put, not only to protect new parts, but also to repair and extend the life of worn parts. Much of the growing use of surfacing has resulted from the need to find ways to extend plant life.

Surface coating technology has developed its own jargon and a glossary of terms is included to minimise the risk of misunderstanding.

Mechanisms of wear

CHAPTER 1

The aim of this chapter is to describe the processes by which wear occurs and to indicate properties needed in materials that are required to operate in wear-producing environments.

Wear may be defined as the progressive loss of substance from the operating surface of a body occurring as a result of relative motion of the surface with respect to another body. The concept embraces metal to metal, metal to other solids and metal to fluid contact and the definition clearly associates the process with the surfaces of materials.

Traditionally, wear has been described in terms of adhesion and abrasion. The former occurs when two bodies slide over each other and surface forces cause the transfer of fragments of material, the latter when particles or protuberances plough or cut fragments from a contacting (softer) surface in relative motion. While abrasion and adhesion are rightly regarded as the predominant wear processes — for example, a survey of industrial wear has shown that they occur in 50 and 15% respectively of all wear situations — it has become apparent that practically every material damage mechanism can contribute to wear; consequently, almost every physical, mechanical and chemical characteristic of a material is liable to affect its wear performance.

In practical situations, more than one type of wear is generally encountered; in addition, one type may predominate for part of the operating cycle and another for the remainder; finally, the products of one type of wear may themselves cause secondary wear by another mechanism. It is also true that wear is a systems phenomenon, the behaviour of the wearing surface being affected, to a greater or lesser extent, by the other parts of the system. These factors combine to make wear a complex subject and while it is convenient to discuss the different processes separately, their interdependence should be borne in mind.

Basic mechanisms of wear

Adhesion, abrasion and contact fatigue are generally regarded as the three basic wear mechanisms that result in material being removed from the surface of a component; other processes — such as corrosion and erosion — can affect, and may intensify, damage. Their principal features are described below and summarised in the tinted panels.

ABRASIVE WEAR
Abrasive wear arises from the penetration of one surface by a harder body or surface; damage involves a cutting or ploughing action. It may involve particles moving over a surface (two body abrasion), hard particles moving over two moving surfaces (three body abrasion) or a roughened surface moving over and penetrating an opposing surface.

In practice, abrasion is often considered under separate headings, such as gouging, scoring, low stress, high stress, etc, Fig.1.1, but these identify the severity, rather than the type, of wear; in all the basic abrasive action is similar.

Where surfaces are subjected to such conditions, the majority of particles cause little damage, simply rolling or sliding over the surface; it is only those whose attack angle lies between ~80-120° which give rise to significant wear damage.

Hardness and wear
It is possible to think that wear rate is inversely related to hardness. However, practical results on abrasive wear tend not to confirm this direct proportionality. In materials of simple microstructure, there may be a direct relation between hardness and wear rate; this has been shown, for example, for commercially pure metals. However, with materials of more complex structure (typified by most engineering alloys), this is not so. In steels, the

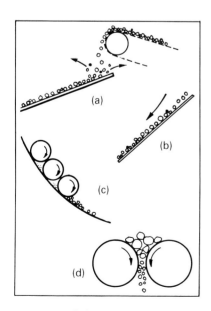

1.1 Types of abrasive wear:
a) Gouging; b) Low stress abrasion; c) High stress abrasion (e.g. rod and ball mills); d) Three body abrasion.

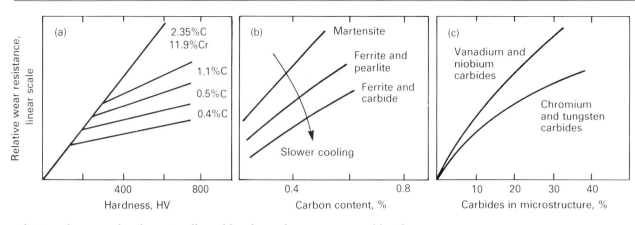

1.2 *Relative wear resistance of steel against:* a) *Hardness;* b) *Carbon content;* c) *Carbides.*

relation of wear to hardness is affected by the carbon content and by the structural condition of the matrix, Fig.1.2.

The presence of secondary phases in the structure is also important. Carbides especially, but also borides, nitrides and phosphides, are widely used with success to provide resistance to abrasion; the degree of improvement depends upon the composition, amount and morphology of the hard phases (as well as upon the operating environment).

With many materials, it is possible to reach a specified hardness by different means. Thus, improvement in resistance of steel to abrasion can be much greater if the matrix is hardened by alloying than if either quenching and tempering or precipitation hardening are used to achieve the same result, Fig.1.3. Again, in cast iron, if the matrix structure is kept constant and the hardness altered by changing graphite distribution, wear resistance can improve with reduction in bulk hardness of the alloy. While these results refer only to the particular test conditions used, they do illustrate that not only hardness but also the method of achieving it are important.

In considering hardness, the differential between the abrading body and the other surface is important. Table 1.1 lists typical hardness data. Wear of a surface tends to be progressively reduced if the ratio of its hardness to that

Table 1.1 Hardnesses of some abrasives and mineral phases

Raw material/mineral hardness, HV		Material phase hardness, HV	
Coal	32	Ferrite	70-200
Gypsum	36	Pearlite	250-460
Lime	110	Austenite	170-350
Calcite	140	Martensite	500-1000
Fluorspar	140	Basalt	700-800
Coke	200	Cementite	840-1100
Iron ore	470	Chromium carbides	1200-1800
Glass	500	Alumina	2000
Feldspar	600-750	Niobium carbide	2000
Sinter	770	Tungsten carbide	2400
Quartz	900-1280	Silicon carbide	2600
Corundum	1800	Vanadium carbide	2800
		Boron carbide	3700
		Diamond	10000

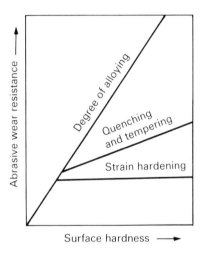

1.3 *Wear resistance for various methods of increasing hardness.*

of the abrading body increases over the range 0.5-1.3. However, this is complicated by the fact that the wear rate/hardness curve generally shows discontinuities, Fig.1.4, the transitions depending upon the hardness of the abrading phases.

Most metallic surfaces work harden during wear. Discussions of the role of hardness have generally centred on the hardness measured on unworn samples, whereas the relevant hardness is that after wear. These values can be quite different, the most significant example being austenitic manganese steel which, from an initial hardness of 200HV, can be work hardened to 600HV. Indeed, this material only gives good wear resistance to abrasive conditions if there is sufficient impact loading to ensure transformation of austenite to martensite during operation.

These observations should not be taken to indicate that hardness — even as usually measured on unworn components — is unimportant but rather that

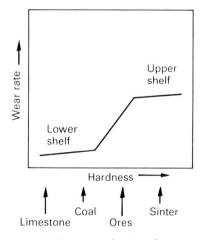

1.4 *Wear rate of material versus hardness of minerals.*

it alone does not determine the ability of a material to resist abrasion; other properties also have an influence. Structure and composition play an important part; differences in manufacture can also influence wear resistance, via their effects on material structure.

Control of abrasive wear

In seeking abrasion resistant materials, it is necessary to ensure there is sufficient strength to resist imposed stresses from the wear mechanism and sufficient toughness to withstand impact. For pure abrasion, the ideal structure tends to be a hard matrix (e.g. martensite) with evenly distributed hard particles (e.g. carbides). If there is a large amount of impact loading, then it may be necessary to revert to a tougher matrix (e.g. bainite) at the expense of some wear resistance, or possibly use an initially relatively soft matrix that transforms during service (e.g. unstable austenite). This dual requirement for strength and toughness demands a compromise. Where there is any doubt, the balance should be tilted in favour of extra toughness; a higher than optimum wear rate reduces component life but less disastrously than component fracture.

ABRASIVE WEAR
- Abrasion involves penetration of one surface by a harder body or surface.
- Abrasion is often described by different terms — such as gouging, scoring, *etc*, but these identify the severity of wear; in all of these, the basic mechanism of damage is similar.
- In general, wear resistance is not related to hardness in a simple way; other factors, particularly composition and structure, markedly influence resistance to abrasion.
- In choosing wear resistant materials, both strength and toughness are normally required and this involves some degree of compromise in selection.

ADHESIVE WEAR

The established theory of adhesive wear is based on the premise that surfaces of crystalline (metallic) solids, although apparently smooth, are rough on a microscopic scale and consist of a series of peaks and valleys. When two such surfaces are brought together, they make contact only at opposing asperities, Fig.1.5. The area of these asperity contacts may be as little as 10^{-4} times that of the nominal contact area; therefore, even a modest load applied normally to the surfaces causes high local pressures (and temperatures). The elastic limit of one (or both) materials may be exceeded and the asperity contacts undergo plastic deformation, until the real contact area has increased sufficiently to support the applied load. Plastic flow disrupts protective surface films and so allows clean metals to come into intimate contact; this may cause welding and the junctions so formed tend to rupture upon relative movement of the surfaces. As sliding proceeds, this sequence of asperity welding and junction rupture is repeated. If rupture takes place on the original joint interface, minimal damage is caused; however, if the welds work harden (as is common with metals), rupture is more likely to occur behind the interface and fragments will be torn out of one (or both) of the contacting surfaces, which thereby suffer damage. In addition, the loose debris produced may itself contribute to wear by abrasion.

It has been claimed that friction and wear are directly related, but as Table 1.2 shows, this need not be so; high friction may be associated with low wear and vice versa. Neither friction nor wear is an intrinsic property of a material and their values are determined by the engineering system. It is therefore unwise to attempt to extrapolate wear performance from friction results; at best, changes in coefficient friction within a single wear couple may be taken to indicate possible changes in the dominant wear mechanism.

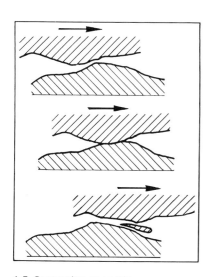

1.5 *Contacting asperities.*

Table 1.2 Wear and friction data for different materials

Materials	Wear rate, cm^3/cm	Coefficient of friction
Mild steel on mild steel	1.57×10^{-7}	0.62
60/40 leaded brass	2.4×10^{-8}	0.24
PTFE	2.0×10^{-9}	0.18
Stellite	3.2×10^{-10}	0.60
Ferritic stainless steel	2.7×10^{-10}	0.53
Polyethylene	3.0×10^{-11}	0.65
Tungsten carbide on tungsten	2.0×10^{-12}	0.35

Note: Load 400g; speed 180 cm/sec. Materials run against hardened tool steel, except for mild steel and tungsten carbide.

Mild and severe wear

Adhesive wear is noted for marked transitions in wear rate resulting from changes in load or sliding speed; this is a consequence of the fact that there are two different regimes of damage, known as mild and severe wear. Figure 1.6 illustrates such transitions for steel. Below T1, the wear rate is low and the process conforms to mild wear; at T2, the wear rate increases greatly as severe wear begins and this high rate continues until it suddenly drops at T2 to another period of mild wear.

Mild wear depends upon the presence of an oxidised surface which keeps clean metals apart and the wear process involves the production of fine, non-metallic debris from this surface layer. The protective oxide films are mainly generated during rubbing or sliding, so that conditions that favour oxide formation are most likely to confine wear to the mild regime. Oxide particles that become detached can act on the system either as a lubricant or as an abrasive, depending upon their nature.

When the load on the surfaces becomes sufficiently high, fine cracks begin to form in the oxide layer and these may be followed by plastic flow of the supporting material; this causes the onset of the severe wear regime. There is significant material transfer from one surface to the other, metallic fragments are torn out and damage may be as much as 10 000 times as great as in mild wear. There are at least two distinctive stages in the process: the removal of metal fragments from the wearing surfaces and the formation of wear particles which may be 5-10 times as large. There may also be a third stage, in which the transferred material is oxidised.

During sliding under dry or imperfectly lubricated conditions, high temperatures can be generated by friction and these can lead to another transition in wear rate. High temperatures may cause intense surface hardening and possibly structural transformation in the sliding members. In addition, the elevated temperatures may affect the nature of the oxide formed on the surfaces, often to a more protective variety. The consequence of these changes is frequently a material more resistant to deformation and on which a protective surface layer can be re-established. If this happens, the wear rate drops again, to a level consistent with mild wear.

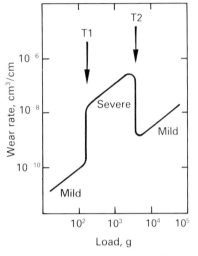

1.6 Transition wear behaviour for steel.

Running-in wear

During the initial stages of sliding, particularly with unused components, metal surfaces tend to be clean, with little surface film protection; under these circumstances, special care must be taken to avoid damage. This transient type of condition is known as running-in and it has been observed for many materials under a variety of operating conditions. Running-in may produce few adverse effects, since an oxidised or lubricated surface layer is quickly established and this leads to an acceptable wear rate. However, under certain operating conditions, a form of severe wear known as scuffing can occur. This is characterised by roughening of the surface, beginning in narrow, localised bands and gradually spreading around the rubbing surface. Components seem to be most at risk from scuffing at loads just below the mild/severe transition (Fig.1.7) and under conditions of poor lubrication.

Control of adhesive wear

If direct and intimate contact of opposing surfaces can be prevented, there is no adhesive wear. Surface films, whether of oxides or absorbed atom layers,

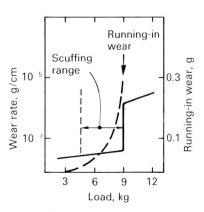

1.7 Scuffing range for grey cast iron against steel.

are extremely important and when these are deliberately removed high wear rates result. Materials with good oxide forming capacity may be expected to perform well but the situation is complicated by the fact that the sliding action tends to disrupt the films.

Materials of high elastic moduli have been claimed to offer advantages in resisting wear, as the area of asperity contact should be limited by the high modulus. However, as with hardness, it is difficult to equate modulus in any simple way with wear performance. Perhaps one of the most important properties is the ability of a material to work harden, as this controls the position at which asperity junction rupture takes place and so affects the degree of damage.

Materials options to control or minimise adhesive wear are, in practice, usually based upon one (or more) of three approaches:

1. Lubrication using oil films to prevent direct metal/metal contact. However, it should be emphasised that it is not possible to maintain full hydrodynamic lubrication at all times and some contact of mating surfaces is inevitable.

2. Use of materials that are mutually insoluble. This may not always be possible, because of other constraints in choice of materials for opposing surfaces.

3. Application of surface coatings (or treatments). The success of such techniques depends mainly upon their ability to interpose a dissimilar and non-reactive barrier between otherwise reactive surfaces; the porosity inherent in some types of coating can provide a further benefit, by providing a reservoir for oil which can be helpful during periods of temporary lubricant starvation.

> **ADHESIVE WEAR**
> - Adhesive wear is characterised by sharp transitions in behaviour and wear may change from mild to severe (and back again) with only slight changes in load or sliding speed.
> - Adhesion of surfaces can take place and cause significant wear damage only if the asperities of the opposing surfaces make contact metal to metal.
> - For the majority of engineering materials, there is no evidence that hardness can be directly linked to adhesive wear resistance.
> - Control of adhesive wear is usually achieved by one (or a combination) of three methods: lubrication, use of mutually incompatible materials in the wear couple and/or application of surface treatments/coatings that interpose a non-reactive layer between the metallic surfaces in relative motion.

CONTACT FATIGUE

Rolling fatigue

Contact fatigue (also known as surface fatigue) occurs in surfaces in rolling contact, such as ball and roller bearings and gears. During rolling, surfaces are exposed to fatigue conditions and, if the endurance limit is exceeded, fatigue failure eventually occurs; cracks propagate through the material and these may join up to isolate pieces of material that may then fall out. Fracture, under these circumstances, takes place by a normal fatigue process.

Adhesive and abrasive wear involve progressive loss of material from contacting surfaces; if the surfaces are kept apart (and abrasive particles excluded), wear does not occur. Contact fatigue satisfies neither of these criteria. Failure occurs suddenly, after an incubation period during which there is little (if any) loss of performance; direct contact of the opposing surfaces is not necessary, as the stresses can be transmitted through intervening material such as a lubricant film.

The location of the failure should be defined by the position of maximum shear stress, Fig.1.8, but, in practice, it is greatly influenced by the presence

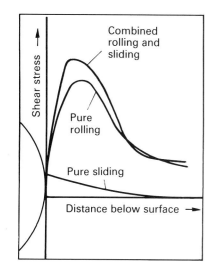

1.8 *Variation of shear stress with distance below surface.*

of defects in the material. Wear caused by contact fatigue is assessed in terms of useful life, defined as the number of revolutions under a given stress that will be exceeded by 90% of the test components.

Contact fatigue can be classed under two headings, depending upon whether cracking has initiated at the surface or within the material. The former tends to occur with hard components in mutual contact in the presence of rolling and sliding. This type of failure is known as pitting and the particles of debris produced are usually triangular in shape. Failure originating below the surface generally stems from voids, inclusions or hard particles. Cracks propagate parallel to, or beneath, the surface and subsidiary cracks appear which, if they join up, can give rise to loose particles that may fall out. With case hardened material, it is not unusual, if there is insufficient depth of hardening, for cracking to be located at the interface between case and core of the component.

Control of rolling fatigue

Possibly the biggest single factor in contact fatigue wear is the load; even a small reduction greatly increases the life of a part. In terms of materials properties, an increase in hardness may increase pitting resistance; further, in respect of surface-initiated fatigue particularly, surface finish is a determining factor. However, too great an increase in hardness can militate against the surface adjustment which is desirable in running-in. Compromise is therefore necessary and this commonly involves manufacturing one of the contacting surfaces slightly softer than the other. With subsurface fatigue, hard materials are again desirable but they become less conformable as hardness increases; yet again, a compromise is effected by making the surface only as hard as is required.

Lubrication has little direct effect on contact fatigue, as the process depends on the stress versus cycles criteria. However, it can help indirectly by keeping surfaces dirt free and smooth, by eliminating adhesive wear and corrosion effects and by smoothing out local stress concentrations.

CONTACT FATIGUE
- Fatigue occurs suddenly, after an incubation period during which there is little (if any) loss of performance; direct contact of the opposing surfaces is not necessary. In these respects, contact fatigue differs from abrasive and adhesive wear.
- Failure may take place either at the surface of the component or within the material. The precise location of failure is greatly influenced by the presence of defects in the components.

Contributory processes and effects

FRETTING

Fretting is a consequence of two surfaces having an oscillatory relative motion of small amplitude; the surfaces do not lose contact with each other and the products of wear are unable to escape, so that seizure may occur because of blockage of the free space and prevention of access for lubricant. Although it is convenient to treat it separately, fretting is not a distinct type of wear but is a form of adhesive wear.

Fretting probably involves three main stages. Firstly, mechanical action disrupts the oxide films on the surfaces and exposes clean, reactive metals; although these may re-oxidise during the next half cycle, they are disrupted again on the return half cycle. The metal particles are then removed from the surface in a finely divided form by mechanical grinding or by formation and breaking of welded asperities. Finally, oxide debris, either from the first process or by oxidation of the metallic particles released in the second, forms an abrasive powder which continues to damage the surface.

Fretting can be removed by eliminating the source of the vibration; alternatively, surrounding the contact area with lubricant reduces friction and the adhesion propensity of the surfaces. With ferrous components particularly, elimination of oxygen is beneficial, as it militates against formation of oxides. Specific fretting difficulties can often be overcome by application of surface coatings and treatments that reduce metallic adhesion, such as phosphiding, have been particularly successful in this respect.

EROSION

Erosive wear is a special form of abrasion in which contact stress arises from the kinetic energy of solid or liquid particles in a fluid stream encountering a surface. The damaged surface may show a fine granular appearance, rather similar to that observed in brittle fracture. Particle velocity and impact angle, together with abrasive size, give a measure of the kinetic energy of the erosive stream, and the wear volume is proportional to the cube of the stream velocity.

In general, there is a correlation between the erosion resistance of a material and its 'ultimate resilience' (defined as $\frac{(UTS)^2}{2E}$

where UTS = ultimate tensile strength
E = modulus of elasticity).

This expression is a measure of the amount of energy that can be absorbed before cracking occurs. If the impact angles in erosion are small, cutting wear is the predominant mode of damage and hardness of the surface is the main requirement; hard materials, even if brittle, are generally satisfactory. With larger impact angles, wear involves deformation and a measure of toughness becomes necessary; soft materials may be suitable and, indeed, rubbers — because of their low moduli of elasticity — are often best, Fig.1.9.

It cannot be overstressed that erosion is often most easily (and economically) overcome by changing the angle of attack, or by reducing the velocity of flow, rather than simply by providing a more erosion resistant material, whether bulk or coating.

Cavitation erosion is a particular form where a solid and fluid are in relative motion and bubbles formed in the fluid become unstable and implode against the surface of the solid. The stability of the bubble depends on the internal and external pressure differential, and on its surface energy which is a measure of the potential damage on collapse. A reduction of surface tension of the liquid, or an increase in vapour pressure, tends to reduce damage.

Another special form of erosion is produced when an electric spark occurs between two surfaces and causes permanent damage; this type of spark erosion produces problems for electrical contacts.

CORROSION

When sliding takes place in a corrosive environment, surface reactions occur and reaction products are formed on one or both surfaces. Various types of reaction product can be formed. In dry air at elevated temperatures, the product is an oxide and in moist air an oxide or hydroxide; with some metals it may be a carbonate; and in industrial atmospheres, chlorides, sulphides and nitrates are common. In all of these, the reaction product tends to adhere poorly to the surface and further rubbing removes it mechanically. In this way, bare metal surfaces are continuously exposed and the medium may appear much more aggressive than it is under static conditions.

This type of reaction is commonly referred to as corrosive wear but it is not a specific form of wear and should more accurately be termed wear affected by corrosion.

The presence of a lubricant can have a beneficial effect on the system for two reasons. Firstly, it may protect surfaces from the corrosive media and reduce frictional heating; secondly, the lubricant itself may react with the surface to produce a different form of reaction product. This latter effect is deliberately promoted with EP additives in oil, in which an active chlorine or sulphur compound is used to form anti-scuff and anti-seizure coatings.

ELEVATED TEMPERATURE

The temperature of the rubbing surface is important in wear and elevated temperatures can have a number of effects on wear performance. Strength and hardness of metallic materials decrease with increasing temperature

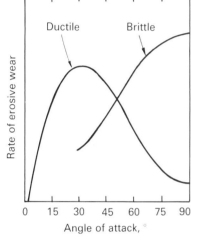

1.9 *Dependence of rate of erosion on angle of attack of impinging particles.*

and the reduction can be rapid as temperatures become high; the relative strength of any adhesive joints formed become relatively much greater under these conditions and this increases the likelihood of surface damage. Materials in the wear couple are often specified to have high hot hardness to counteract these effects; tool steels and alloys based on cobalt, with chromium and molybdenum or tungsten are commonly used. Plastics, and plastics based materials, suffer greater loss of strength properties than metals when heated. Ceramics, and some cermets, retain hardness at high temperature better than metals or plastics and, for highest temperature operation, ceramics (or possibly cermets) may be the most appropriate choice, provided the lack of ductility can be accepted.

Surface contaminating films are affected by temperature. Formation of different types of oxide at different temperatures has already been discussed. Another effect is the removal, by heating, of adsorbed surface films — usually moisture or organics — which tend to prevent severe metallic wear taking place.

Lubricants such as oil suffer degradation at quite moderate temperatures (<300°C) and even synthetic oils cannot be used for lengthy periods at much higher temperatures. Above about 500°C, some solid lubricants may be suitable but, again, their use is limited by chemical factors.

The effects of elevated temperature on wear are generally difficult to predict with accuracy but usually it would be expected to make the wear rate greater.

Testing for wear resistance

Although many wear tests have been carried out on a variety of machines the conclusion which seems to have been established is that equipment should simulate the service conditions of the component as closely as possible. This is understandable if the wide variations in service conditions are considered; for example excavator bucket teeth are subject to non-metallic abrasion at ordinary temperatures, conditions totally different from those arising in, say, the shearing of hot steel billets. Generally, results obtained in a testing machine, which carefully simulates certain specific service conditions, do bear a relationship to service results and the machine can be used to test different alloys to sort them out. For example, rubbing blocks on discs to test screw flight/barrel combinations for plastic processing equipment resulted in marked differences. Three barrel materials: martensitic white iron 58-64RC iron; a corrosion resistant Co-Ni-Cr-B alloy 48-54RC; and a composite of tungsten carbide particles dispersed in a hard, corrosion resistant Ni base matrix were run against four surface treated steels and two surfacing alloys. The results are shown in Table 1.3.

Table 1.3 Block on disc adhesive wear tests for various screw flight/barrel combinations of materials

Barrel material (B)	Martensitic white iron 58-64RC			Corrosion resistant Co-Ni-Cr-B alloy 48-54RC			Composite WC in hard, Ni base matrix		
Screw material (M)	Weight loss (M), mg	Weight loss (B), mg	Scar width, mm	Weight loss (M), mg	Weight loss (B), mg	Scar width, mm	Weight loss (M), mg	Weight loss (B), mg	Scar width, mm
Co-Cr-W-C surfacing alloy (40RC)	7.6	2.4	2.01	7.35	10.2	3.46	Galled and seized		
Ni-Cr-Si-B surfacing alloy (60RC)	1.2	2.3	2.00	0.4	1.5	1.84	2.2	0.2	0.52
Carburised SAE 4620 steel	3.3	1.5	1.83	3.8	0.95	1.56	13.85	0.3	0.83
Hardened SAE 4140 steel	27.05	1.1	1.68	27.1	1.3	1.70	100.4	0.4	0.69
Nitrided SAE 4130 steel	7.35	1.4	1.79	8.0	1.2	1.75	156.7	0.7	1.40
Nitrided Nitralloy	7.0	2.0	2.04	4.05	1.55	1.84	26.6	0.1	0.65

Note: Loads used in the tests represented extreme conditions. There was no interfacial film between barrel and screw materials, as in actual use.

The tests show that there is no direct relationship of wear resistance to material hardness and that the compatibility of any two materials together is not related to their relative hardness, but the data is useful if used judiciously as a guide to the better alternatives. Adhesive wear test results cannot provide the only criterion to material selection — corrosive conditions must be considered, also material strength requirements. Thus, although the case-hardened 4620 steel shows excellent compatability with the first two barrel materials, it may not have sufficient strength for many applications. Past service experience with the Co base alloy (an intuitive choice) has shown results better than indicated, in the majority of cases. Where problems have occurred, the tests can suggest probable improvements.

The harder the surface the better the resistance against abrasion. This is true if one is dealing with monobloc materials, but most alloys contain two or more phases and many of the surfacing alloys have complex structures with hard phases dispersed in a softer matrix. Hardness of constituent parts, their distribution and size have to be considered in relation to the operating regime. Generally, the macrohardness of the material to resist abrasion should be at least 80% of the hardness of the abradent.

In industrial practice greatest resistance to wear has been found in materials containing the maximum amount of the hardest carbides or borides which can be supported adequately in a suitable matrix. The balance between the amount and size of the hard particles and the properties of the matrix must be related to the wear conditions, which may be metal to metal at high or low velocity, fretting, abrasion by lightly or heavily loaded particles of small or large size, impact or erosion by liquid or gas. Any of these factors may be complicated by a hot or corrosive environment.

For many applications involving abrasive wear, tungsten carbide is a preferred material. The greater the percentage of the carbide the better the wear resistance, so massive carbide with 6-25%Co as the binder (percentage adjusted to the degree of toughness needed) or applied as a coating, although expensive, may justify the cost. However, in an application in which the abrasive is fine, performance may be bad and far below a coating with only perhaps 30-50% tungsten carbide in a matrix of an Ni base alloy. This is because the soft Co phase may be eroded out leaving the unsupported carbide to fall out while the harder Ni alloy resists this erosion better.

Often trials on the plant must be made to determine the best surfacing solution. Results reported by an iron and steel company on protection of breaker tips in the sinter plant show the improvements achieved. The chosen method involved use of chromium and tungsten carbides with a life about double that of the minimum tolerable intervals between maintenance schedules. To obtain maximum wear resistance, tubular tungsten carbide rods were deposited by gas welding as this minimises solution of carbide in the steel matrix alloy compared with arc welding and this gives maximum wear life. With the introduction of a high percentage of foreign ore into the sinter mix, wear increased considerably. To reach tolerable maintenance levels extensive use of tungsten carbide became essential, even then results were not as good as with the less abrasive home ore, see Table 1.4.

Table 1.4 Typical maintenance levels required by hard surfaced sinter plant

Sinter practice	Breaker tip type	Service life		Initial cost of unit, £	Cost/tonne sinter, p
		Months	Sinter output, t		
Home ore	26-28%Cr tip	18	1.25M	2500	0.20
Foreign ore	26-28%Cr tip	9	600 000	2500	0.42
	Hardfaced with high tungsten carbide	12-14	900 000	4500	0.50

Another method of deposition was used for tests on nozzle guide vanes which suffered severe wear in aeroengines flying around Iceland where the atmosphere contains much volcanic ash. Detonation gun deposited tungsten carbide test areas showed excellent resistance so adequate surface protection is now designed into the wear vulnerable sections. This emphasises the desirability of examining wear patterns as part of the design process. With the work referred to above, the more expensive tungsten carbide was applied only to the areas experiencing the greatest wear.

Practical diagnosis of wear

Earlier sections of this chapter deal with the theory of wear. It is important, when examining a worn part before choosing a surfacing process and material to give improved performance, to identify which wear mechanisms are at work. This then forms the basis for making that choice.

Wear mechanisms identified earlier in this chapter are adhesion, abrasion, contact fatigue, fretting, erosion, corrosion and elevated temperature. In most applications, the appearance of the worn surface will suggest the dominance of particular mechanisms. However, the ability of others to accelerate the main forms of attack, without necessarily leaving obvious evidence of their influence means that careful consideration of the possibilities must always temper visual evidence.

ABRASION

The appearance of abrasive wear is influenced by many factors including the materials involved, pressure and speed of rubbing impact, turbulence of movement, *etc*. As a result the appearance of a metal surface exposed to wear by abrasive materials is usually characterised by regular or irregular scoring, scratching and pitting.

The presence of corrosion may not be apparent on the abraded surface because of the continual cleaning effect of the abrasive but there may be signs of corrosion products in any stagnant areas of movement. The metal may claim good resistance to the medium/media involved, but if this depends on formation of protective surface films, the abrasive will remove these and continuously expose clean areas to the combined effects of both, as indicated earlier in this chapter.

If three body abrasion is involved, for example a metal to metal bearing which is exposed to abrasive material, there is the possibility that abrasive particles will be pressed into the surface of the softer of the two materials and act as a lap. After a period, the lap shows little or no signs of wear while the harder material is significantly worn.

ADHESION

Adhesion occurs in metal to metal rubbing and is usually characterised by transfer of metal from place to place on the surface or from one surface to the other. Variations in fit, alignment and rubbing speed and load may cause a rubbing pair to move into an adhesive wear mode. Its occurrence can also be influenced by the presence of oxides or corrosion products as indicated earlier. Adhesive wear is responsible for the well known and characteristic condition of seizure which can occur in closely fitting rubbing metal surfaces, but if clearance is sufficient to expel wear debris, wear can continue and result in significant loss of size to the surfaces involved. This is evident in Fig.1.10, which shows the worn pin of a link chain.

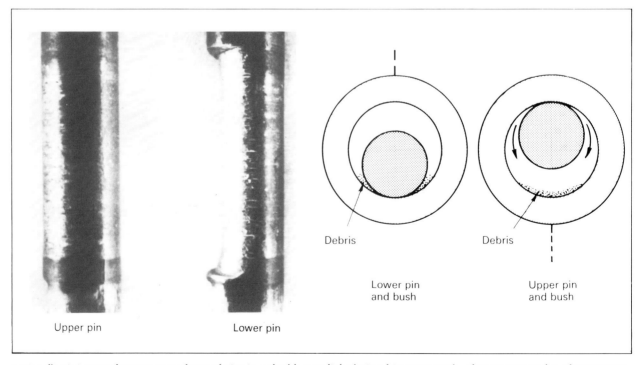

1.10 *Adhesive wear damage to surfaces of pins in a double row link chain of 25.4mm pitch, after 11 000 cycles of movement through 90° over a pulley at a frequency of 0.5Hz. Note the difference in wear between the upper and lower pins, because of wear debris trapped between the lower pin and bush (courtesy CEGB).*

1.11 *Fretting wear scars where two 6mm wide piston rings were in nominally static contact with a steel shaft but subject to significant vibration. Adjacent marks were caused by the lands on the ring carrier (courtesy CEGB).*

1.12 *The pattern of erosive wear in a steel pipe bend carrying quartz grains in air at a velocity of 27.4 m/sec (courtesy CEGB).*

Corrosion and temperature can play an important role in increasing or decreasing adhesive wear, depending on the influence of the surface films that they produce.

CONTACT FATIGUE

Surface damage caused by contact fatigue arises from the ability of the material used to withstand the cyclic loads imposed in service. When other wear factors are absent, the solution is either to use a material of improved fatigue properties, or to change the design to reduce the stress involved.

Surfacing materials and processes may offer a solution to this type of problem, but selection may be difficult if the failed part itself is a material with good fatigue properties. The best properties for this purpose, available from surface coatings, are associated with weld surfacing processes using ferrous alloys which rely on martensitic hardening, either during welding or from subsequent heat treatment.

FRETTING

Fretting is a particular form of adhesive wear as indicated earlier. It is often characterised by the presence of metal oxide powder in a dismantled joint and may be associated with significant amounts of wear. Figure 1.11 illustrates wear scars 500μm deep on a steel shaft caused by two 6mm wide piston type seal rings. The adjacent three marks were caused by the lands on the ring carrier. In this example, the assembly was exposed to considerable vibration in an axial flow compressor driven by a 5MW motor. There is no reciprocating or rotary movement between rings and shaft.

EROSION AND CAVITATION

It is important to distinguish between these two forms of attack. Erosion is taken to mean the scouring action caused by high velocity low stress abrasion between solids and liquids (or mixtures) and a metal surface. A good example of this is shown in Fig.1.12, which is the internal surface of a pipe bend exposed to air carrying quartz particles up to 500μm at a velocity of 27.4 m/sec (90 ft/sec).

Erosion of surfaces exposed to high velocity fluids, such as steam, which contain no abrasive particles often exhibit a quite smooth surface either over an extensive area or in a series of eroded grooves.

Cavitation, as described earlier, is a stress generated form of attack and can be characterised by an extremely rough surface, consisting of connected pits produced by fatigue fracture of small particles progressively from the surface.

CORROSION

Apart from the catalytic effects of corrosion on other forms of wear, there are two influences worthy of separate mention. One is the phenomenon of crevice corrosion, when adjacent areas of the same metal are unaffected. This can occur in quite narrow gaps between mating surfaces and could, for example, affect fretting wear. The other is that corrosion fatigue can occur as a result of the stress concentrating effects of surface attack by the corrosive medium.

ELEVATED TEMPERATURE

Elevated temperature can affect wear rates in several ways. Hardness and strength of metals and alloys are reduced as the temperature rises, to a greater or lesser extent. These effects may be reversible, *i.e.* on cooling the properties recover, or permanent. Corrosion and oxidation are likely to be accelerated as temperature rises, but as noted earlier, this usually increases the rate of abrasive wear but can reduce adhesive wear if the surface films act as a form of lubricant and prevent metal to metal contact.

These then are some of the factors to consider when assessing characteristics required from a surface coating for a given set of service conditions.

STRIP CLADDING – A NEW DIMENSION TO AN EXISTING PROCESS

Mixalloy manufactures strip metals and alloys from powder and have developed a range of consumables for strip cladding, utilising submerged arc or electro-slag welding techniques.

Mixalloy's established expertise in powder metallurgy provides homogeneity, versatility of composition and high manufacturing standards to ensure a quality range of welding consumables, including iron based - **FEROMIX**, nickel – **NICROMIX**, and cobalt based – **STELMIX**.

Mixalloy's products provide a cost effective method of weld cladding.

This flexible process allows small quantities to individual specifications to be produced in short lead times.

Mixalloy's technical service meets users needs in developing products for specialist applications.

Contact Lawton Fage for further information.

MIXALLOY

Mixalloy Ltd., Antelope Industrial Estate, Rhydymwyn, Mold, Clwyd CH7 5JH. Tel: (0352) 741517 Fax: (0352) 741035 Telex: 617142 Mixloy G

Feromix, Stelmix and Mixalloy are Registered Trademarks of Mixalloy Ltd. Nicromix is a Trademark of Mixalloy Ltd.

Weld surfacing materials and processes

CHAPTER 2

Characteristics of weld surfacing

The principal characteristics of welding processes which distinguish them from other methods of surface coating are as follows.

COATING THICKNESS
Welded deposits of surfacing alloys can be applied in thicknesses greater than most other techniques, typically in the range 3-10mm, with some restrictions for certain materials. This facility is essential when protection in depth is necessary.

ADHESION TO SUBSTRATE
Correctly selected materials and properly operated processes provide a metallurgical bond to the substrate which withstands thermal and mechanical shock without detachment.

RANGE OF PROCESSES
Most welding processes are used for application of surface coatings bringing the operation within the scope of most engineering activities and enabling on-site work to be carried out under certain circumstances.

RESURFACING
The opportunity to carry out repairs on worn parts, whether previously weld surfaced or not, is a feature exploited in several industries.

Welding processes involve application of some heat to the component being processed and depending on the material from which it is made and its condition, certain precautions may need to be taken; these are described in detail in this chapter.

During weld surfacing, the coating material is raised to its melting point which means that metals and alloys used for the purpose must have a melting point similar to or less than that of the steel substrate materials in common use. Other coating materials with higher melting points, such as ceramics, are applied by thermal spraying processes which are described in the following chapter.

Materials for weld surfacing and their selection

Most materials used for weld surfacing have hardnesses greater than HV200 and for this reason they are often referred to as hardfacing or hardsurfacing alloys. Because of the large number of these materials it is convenient to classify them into groups. Such a system developed by the British Steel Corporation is shown in Table 2.1, which gives details of composition and typical uses for each type.

The table included in the introduction to this book is based on the BSC system, but is in a simpler form and incorporates surfacing materials which can only be applied by other surfacing processes. This has been done to simplify the selection of a material to suit given service conditions and to identify the application processes for which it is suited.

GROUP 1 MATERIALS
Carbon steels
Plain carbon steel containing up to 0.5% carbon can be deposited by gas or arc welding processes to produce a weld deposit of about 240HV (20RC). The deposit can be subsequently heat treated to higher hardness if required. The main application of this type of alloy is building up components for subsequent flame hardening or for use as a buffer layer between a softer parent metal and a harder surfacing alloy. Few manufacturers of consumables supply this type of filler metal because its abrasion resistance is low and buffer layers can be made with hydrogen controlled basic covered electrodes which are readily available.

Table 2.1 Consumable classification (after British Steel Corporation)

Classification		Typical composition, %												Notes	Deposited hardness Vickers, HV	Characteristics applicable to deposit	Typical applications
Group	Type	Fe	C	Cr	Mn	Mo	V	W	Co	Ni	B	Nb	Cu				
1 Steels	1 Carbon steels	Bal	1.35												Up to 250 responds to heat treatment	Tough, crack free, machinable, low wear but good impact resistance. Maximum build-up and buttering layers, by gas or electric arc welding	Repair of steel components build-up, alternate layers in laminated surfaces
	2 Low alloy steels	Bal	0.1 0.5	√	√	√	√	√		√	√			Alloys indicated thus total up to 12%	250-650 responds to heat treatment	Properties depend on composition: wide range mainly martensitic structure. Higher alloys: harder and more wear resistant, but lower alloys give toughness and impact resistance — generally machinable — low crack susceptibility may require pre- and postheat treatment	Punches, dies, earth-moving equipment, gear teeth, railway points
	3 Martensitic chromium steels	Bal	0.1 1.7	10 15	√	√	√	√		√				Alloys indicated thus total up to 10%	350-650 responds to heat treatment	Improved wear resistance over class 1/2 with increased oxidation and corrosion resistance. Medium impact resistance decreasing with high C types. May require preheat	Suitable for metal to metal wear up to 600°C. High C types suitable for hot work tool applications. Shear blades, mill rolls, roll necks, crane wheels, hot work dies, and punches
	4 High speed steels	Bal	0.3 1.5	10 max.	5	10	3	20	12		3				600-750 hot hardness to 600°C	Higher carbide types provide good wear resistance, moderate impact resistance. Suitable for elevated temperatures. Lower carbide types for higher impact. Generally grindable; anneal to machine	Hot work dies, punches, shear blades, ingot tongs
	5 Austenitic stainless steels	Bal	0.07 0.2	17 32	√	√				7 22				Mo and Mn total up to 10%	As-deposited 200 up to 500 on work-hardening	Tough, high corrosion, and heat resistant with low abrasive wear resistance. Impact resistance lower than class 1/6	Used as ductile buttering layer, e.g. when depositing high Mn steels on to carbon steel base to avoid brittle bend zone. Furnace parts, chemical plant
	6 Austenitic manganese steels	Bal	0.5 1.0	√	11 16	√	√			√				Cr, Ni, Mo and V total up to 10%	As-deposited 200 up to 600 on work-hardening	Tough, impact resistant: work hardens under heavy impact. Base metal cooling necessary during welding to reduce likely carbide embrittlement. Working temperatures not to exceed 200°C. Buttering layer may be desirable on carbon steel base metals. Arc weld applications only	Heavy impact applications. Crusher and excavator equipment, railway points and crossings, crusher hammers
	7 Austenitic chromium-manganese steels	Bal	0.3 0.5	12 15	12 15	√	√			√				Mo, V and Ni total up to 4%	As-deposited 200 up to 600 on work-hardening	Similar to class 1/6 but high chromium inhibits carbide embrittlement, thus can be deposited on to carbon steels; not restricted to 200°C working temperatures. More abrasion resistant than class 1/6	Crusher and excavator equipment, railway points and hammers
2 Irons	1 Austenitic irons	Bal	4	12 20	√	√				√					300-600 responds to heat treatment	High abrasion resistance, moderate impact resistance. Grindable; pre- and postheat to reduce cracking	Buttering layer to chrome irons. Crushing equipment, pump casings, impellers, excavator teeth
	2 Martensitic irons	Bal	1 4	1 10	√	√	√	√		√		√		Mn, Mo, V, W, Ni, Nb, total up to 25%	500-750 responds to heat treatment	High abrasive wear resistance, low to moderate impact resistance (Ni: hard type). Nb improves wear resistance in hot conditions up to 400°C	Suitable for conformal contacts in adhesive wear situations. Scrapers, bucket tips, forming rolls, cutting tools
	3 High chromium austenitic irons	Bal	3 6	20 40	√	√				√				V, W, Co, Nb for special high temperature applications	500-750 responds to heat treatment	Very high abrasion resistance, low to moderate impact resistance. Pre- and postheat treatment to reduce cracking. Grindable. Oxidation resistant	Shovel teeth, screen plates, grizzly bars, bucket lips
	4 High chromium martensitic irons	Bal	2.5 4.5	20 30	2 max.										500-750 responds to heat treatment	Very high abrasion resistance, low impact. Good hot hardness and oxidation resistance. Susceptible to cracking; reduced with pre- and postheat treatment	Ball mill liners, scrapers, screens, impellers
	5 High chromium complex irons	Bal	2 5	20 40		√	√	√		√	√				600-800 responds to heat treatment hot hardness to 600°C	Extremely abrasion resistant with low impact resistance; latter improved by addition of Nb. Good oxidation resistance and hot hardness to over 600°C with Mo and Co. Pre- and postheat treatment to reduce cracking. Machine by grinding	Hot wear applications, e.g. screens, scrapers, pulverisers, sinter plant
3 Nickel alloys	1 Nickel	8	2		1					85 min.					160	Preheat 150-300°C may be necessary for thicker sections. Peening of deposit for stress relief. Slow cooling necessary — avoid excessive local heat — soft, machinable tough deposit, suitable for thin sections	Classification 3/1, 3/2 and 3/3. General purpose build-up buttering layer for cast irons. Cylindrical blocks, gearboxes, pump housings.
	2 Nickel-copper	3 6	0.35 0.55		2.5					60 70			25 30		130	Preheat 150-300°C may be necessary for thicker sections. Peening of deposit for stress relief. Softer heat-affected zone — minimum residual stresses — easily machinable: strength between classes 3/1 and 3/3	
	3 Nickel-iron	Bal	2		1					45 60					200	Preheat 150-300°C may be necessary for thicker sections. Peening of deposit for stress relief. High strength and toughness. Easily machinable	

√ indicates possible presence of significant constituents

Table 2.1 continued Consumable classification (after British Steel Corporation)

| Classification | | Typical composition, % | | | | | | | | | | | | Notes | Deposited hardness Vickers, HV | Characteristics applicable to deposit | Typical applications |
Group	Type	Fe	C	Cr	Mn	Mo	V	W	Co	Ni	B	Nb	Cu				
3 Nickel alloys continued	4 Nickel-molybdenum-chromium-tungsten	6	2.5	30 max.		17		15	12	Bal		7		Fe, C, Mo, W, Co, Nb are maximum but will vary	250-500	Excellent resistance to corrosion, erosion and oxidation by hot gases. Tough, good impact, and thermal shock resistance. Hot hardness up 500°C. Wear resistance increases with alloy content	Blast furnace bell and hopper seals, valves, dies, chemical plant
	5 Nickel-chromium boron	√	√	5 25		√		√		Bal	1 5			Fe, C, Mo, Si total up to 10%	200-750	Resistant to abrasion, erosion and corrosion and metal to metal wear at elevated temperatures. Additions of W improve high temperature properties	Valves, seating rings, forge dies, pump shafts, chemical plant
	6 Nickel-molybdenum-iron	5 20				20 30				Bal					200-300	Primarily corrosion resistant to HCl, salt spray, and alkalis	Pumps, valves, chemical plant
4 Cobalt alloys	1 Cobalt-chromium-tungsten low alloy		0.7 1.4	25 32				3 6	Bal						350-400	Abrasion, erosion, corrosion and combinations of these factors at elevated temperatures. Hardness retained at high temperature. Manufacturer of specialist guidance necessary on selection	Valve coatings, pump shafts, sleeves. Wear rings, hot shear blades, ingot tongs, bits, dies. Steelworks mill equipment
	2 Cobalt-chromium-tungsten medium alloy		1.0 1.7	25 32				7 10	Bal						400-500		
	3 Cobalt-chromium-tungsten high alloy		1.7 3.0	25 35				11 20	Bal						550-650		
	4 Cobalt-chromium-tungsten-nickel alloys		1.2 2.0	20 25				10 15	Bal	20 25							
5 Copper alloys	1 Copper, zinc, brasses	Up to 40%Zn, balance Co														Good resistance to adhesive wear, anti-seizing properties, and reasonable corrosion resistance	Bearings, slideways, gears, shafts, propellers
	2 Copper, silicon, silicon bronzes	Up to 4%Si, balance Cu															
	3 Copper aluminium, aluminium bronzes	8-15%Al, balance Cu														Primarily intended for lubricated bearing applications	
	4 Copper, tin, phosphor bronzes	4-12%Sn, balance Cu															
6 Tungsten carbide	1 +10 mesh (+1.7mm)													Minimum of 40% tungsten carbide normally in an iron base matrix but can be in a copper or cobalt matrix. Classification is according to the tungsten carbide particle size. The designation −10 +20 mesh means material which will pass through a sieve of 10 mesh and does not pass through a sieve of 20 mesh. Figures in parenthesis are the corresponding nominal aperture size of sieve in accordance with BS 410		Extreme wear resistance and impact resistance dependent on particle size range, dispersion, and matrix hardness	Sinter plant high wear areas; main suction fan impellers, sinter breakers, etc. Crusher hammers, scrapers, various
	2 −10 +20 mesh (−1.77mm +745µm)																
	3 −20 +40 mesh (−745 +150µm)																
	4 −40 +100 mesh (−390 +150µm)																
	5 −100 mesh −150µm																
7 Chromium-boron paste	1			80						20					750-850	Fine mesh powder in suitable binder to form a paste. Applied to steel base material and fused by carbon arc or gas weld technique. Very hard wear resistant surfaces, low impact resistance	Medium abrasive sliding wear situations. Chutes, selected in blade applications

√ indicates possible presence of significant constituents

Characteristics
High tensile and compressive strength. High impact strength. Low abrasion resistance. Low resistance to tempering. Maximum service temperature 200°C.

Applications
Building up gear teeth, *etc*, for subsequent flame hardening. Buffer layers.

Low alloy steels
Martensitic alloy steels are the most widely used hardfacing alloys and are characterised by low cost and a wide range of properties depending on composition. They contain, in addition to carbon, varying amounts of chromium, manganese, molybdenum, and nickel, as well as smaller additions of tungsten and vanadium.

Characteristics
Hardness 250-800HV (20-62RC). Wide range of combinations of abrasion and impact resistance. Some alloys are heat treatable, *i.e.* can be hardened for increased wear resistance or softened for machining.

Applications
Agricultural implements, earthmoving equipment, crushing machinery, gear teeth, railway points and crossings.

Martensitic chromium steels
Martensitic chromium steels containing about 12% chromium have increased heat and corrosion resistance compared with low alloy steels.

Characteristics
Resistance to metal wear at 300-600°C. Moderate impact resistance.

Applications
Mill rolls, roll necks, crane wheels, hot work tools.

High speed steels
High speed steels are tool steels which are used to cut metals at high rates and withstand temperatures of up to 600°C without softening.

Characteristics
Higher carbon types have good wear resistance and moderate impact resistance. Lower carbon types have higher impact resistance. Heat treatable, *i.e.* can be hardened for increased wear resistance or softened for machining.

Applications
Cutting tools, hot work, ingot tongs, punches, shear blades.

Austenitic stainless steels
Stainless steels have high corrosion and heat resistance, but an important use is buttering carbon or alloy steels before manganese steel is deposited. This avoids formation of brittle phases which occur at the interface between manganese steel and carbon or low alloy steels. Austenitic stainless steels withstand oxidation at temperatures of between 450-600°C, depending on composition. Room temperature hardness is low and thin steels are unsuitable for abrasion resistance.

Characteristics
Tough, but lower impact resistance than austenitic manganese steel. Oxidation resistance up to 600°C, depending on composition. As-deposited hardness, 200HV, which increases by work hardening to 500HV.

Applications
Chemical plant, furnace parts, buttering layer on carbon alloy steels before deposition of manganese steel.

Austenitic manganese steel
Weld metal containing 12-14% manganese has an austenitic structure which is soft (200HV) but which work hardens at the surface to about 600HV (54RC) under heavy impact. In the soft as-deposited condition there is little resistance to low stress scratching abrasion: thus, for example, the weld metal is steadily lapped away when digging in sandy soil. However, if impact is involved, as with rock crushing hammers, the weld metal deforms and work hardens at the surface giving high resistance to further impact. This type of deposit also work hardens if used for digging in soil which contains boulders, and the surface develops resistance to abrasion by soil particles.

Characteristics
High work hardening capacity and extremely high impact resistance. Becomes embrittled if heated above 400°C. Not suitable for deposition by oxyacetylene welding because of embrittlement on slow cooling. Resistance to embrittlement by heat increased by addition of 3-5% nickel or 0.5-1.5% molybdenum. Molybdenum-containing deposits have higher yield strength which gives greater resistance to deformation. Dilution by carbon or alloy steel parent metal produces brittle interface; avoided by buttering layer of stainless steel.

Applications
Rock crushers, pulverising hammers, railway points and crossings, excavating equipment in rocky soil.

Austenitic chromium-manganese steel
Austenitic chromium-manganese steel is used for similar applications to austenitic manganese steel but has certain advantages in spite of the higher cost. Because of the high alloy content these electrodes can be used to weld directly on to carbon steel parent metal without formation of a brittle martensitic interface, and also to weld manganese steel inserts directly to carbon steel.

Characteristics
High work hardening capacity and extremely high impact resistance. Can be deposited directly on to carbon or low alloy steel without the need for a stainless steel buttering layer. Because of high chromium content weld metal cannot be cut or gouged with a gas flame.

Applications
Rock crushers, pulverising hammers, railway points and crossings, excavating equipment in rocky soil.

GROUP 2 MATERIALS
Austenitic, martensitic, high chromium and complex irons
High chromium, austenitic, martensitic, and complex irons often referred to as 'chromium carbide types' contain about 30% chromium and the microstructure of the deposit consists of chromium carbides in a matrix which can be austenite, martensite, or a mixture of both depending on the composition. These alloys are available in the form of cast rods or steel tubes containing chromium carbide particles, and can also be produced as a chromium steel core wire with alloying additions included in the flux covering. Coils of tubular flux-cored wires are also available for semi-automatic or fully mechanised welding.

Characteristics — austenitic irons
High abrasion resistance. Moderate impact resistance. Non-machinable.

Applications
Crushing equipment. Brick, cement, and clay processing equipment. Impellers, excavator buckets and teeth.

Characteristics — martensitic irons
Higher abrasion resistance than austenitic iron. Low to moderate impact resistance. Non-machinable.

Applications
Crushing equipment. Brick, cement and clay processing equipment. Excavator buckets.

Characteristics — high chromium austenitic irons
High abrasion resistance under low stresses. Low to moderate impact resistance. Oxidation resistant. Hot hardness up to 450°C.

Applications
Agricultural machinery in sandy soil, chutes.

Characteristics — high chromium martensitic irons
High abrasion resistance under low or high stresses. Good impact resistance if well supported by rigid parent metal. Oxidation resistant. Heat treatable. Hot hardness up to 450°C.

Applications
Ball mill liners, impellers for gravel, dredging equipment, mine and quarry screens, sand blasting plant.

Characteristics — high chromium complex irons
High abrasion resistance under low or high stress. Good impact resistance if well supported by rigid parent metal. Oxidation resistance. Maintains hot hardness up to 600°C.

Applications
Hot wear conditions, *e.g.* sintering plant, chutes, pulverisers, screens, steel mill guides.

GROUP 3 MATERIALS

Nickel alloys
Pure nickel, nickel-copper and nickel-iron alloys listed are designed to weld cast irons.

Characteristics — nickel
Soft and machinable.

Applications
Surfacing cast iron, buttering layer on cast iron.

Characteristics — nickel-copper
Soft and machinable.

Applications
Surfacing cast iron.

Characteristics — nickel-iron
Machinability lower than nickel or nickel-copper but improved by preheat of 150-300°C.

Applications
Surfacing cast iron, buttering layer on cast iron.

Nickel-molybdenum-chromium-tungsten
Nickel-molybdenum-chromium-tungsten alloys are used primarily for corrosion resistance but they also have good heat resistance.

Characteristics
Good corrosion and heat resistance. Good resistance to erosion by hot furnace gases carrying abrasive particles. Good impact resistance. Hot hardness up to 500°C.

Applications
Blast furnace bell and hopper seals. Hot work tools, *e.g.* die blocks.

Nickel-chromium-boron
The most common nickel based hardfacing alloys contain chromium, boron, and carbon, and their microstructure consists of wear resistant chromium carbides and borides in a nickel-chromium matrix.

Characteristics
Oxidation resistance up to 850°C. Hot hardness up to 500°C. Good corrosion resistance against steam, salt water, and salt spray. Low impact resistance.

Applications
Valves, seating rings, screw conveyors, pump shafts, chemical plant.

Nickel-molybdenum-iron

Characteristics
Resistant to corrosion by HCl, salt spray and alkalis.

Applications
Pumps, valves, chemical plant.

GROUP 4 MATERIALS
Cobalt alloys
The most widely used hardfacing alloys in the non-ferrous group are those based on the ternary system cobalt-chromium-tungsten developed under the trade name of Stellite. Their composition endows the alloys with hot hardness so that they can be used at service temperatures above 600°C with minimum softening or deformation. The presence of chromium forms a closely adherent oxide film which provides oxidation resistance, and this element, together with tungsten, brings about appreciable secondary hardening by the precipitation of carbides. The Stellite alloys have good resistance to erosion, cavitation and adhesive wear and these properties combined with good corrosion resistance enable them to resist the combined effects of different wear types.

Characteristics — cobalt-chromium-tungsten low alloy
Resistant to heat, corrosion, and oxidation under impact stresses.

Applications
Exhaust valves for petrol and diesel engines. Fluid flow control valves, hot metal working tools.

Characteristics — cobalt-chromium-tungsten medium alloy
Higher abrasion resistance than low alloy type but lower resistance to impact or thermal shock.

Applications
Knives used in paper, carpet, and chemical industries.

Characteristics — cobalt-chromium-tungsten high alloy
Highest abrasion resistance and lowest shock resistance of this group.

Applications
Pumps, bearing surfaces, seals.

Characteristics — cobalt-chromium-tungsten-nickel alloy
High erosion and corrosion resistance at elevated temperatures. High thermal shock resistance.

Applications
Exhaust valves for petrol engines.

GROUP 5 MATERIALS
Copper alloys
Copper-aluminium alloys are used for bearing, corrosion resistant, and wear resistant surfaces. Tin bronzes are used for bearing surfaces and occasionally in corrosion resistant applications. Silicon bronze is used for corrosion resistance only and brasses are used occasionally for bearing surfaces.

Characteristics — brasses
Low hardness. Low abrasion resistance.

Applications
Limited use for bearing applications.

Characteristics — silicon bronzes
Good corrosion resistance.

Applications
Overlaying silicon brass.

Characteristics — aluminium bronzes
Good bearing properties. Good corrosion resistance.

Applications
Bearing surfaces deposited on steel, slideways, wear plates.

Characteristics — tin bronzes
Good bearing properties.

Applications
Bearing surfaces deposited on steel, slideways.

GROUP 6 MATERIALS
Tungsten carbide
Tungsten carbide has extremely good wear resistance. Hardfacing rods are actually steel tubes containing tungsten carbide particles (both WC and W_2C) with the relative amount of tungsten carbide to steel at about 60:40. The carbide particles vary from 200 to 8 mesh (0.002 to 3mm), but surfacing rods generally contain different ranges of mesh sizes, e.g. 10/20, 20/30, etc, depending on service conditions. Tubes may be bare for oxyacetylene deposition and either bare or covered for arc welding. Rods are also available in sintered form which contain up to 80% tungsten carbide. Coils of tubular flux-cored wires are also available for semi-automatic or fully mechanised welding.

Characteristics
Oxyacetylene or TIG deposits contain unmelted carbides giving extremely high abrasion resistance. Arc welding melts most of the carbide granules giving a homogenous weld having slightly lower abrasion resistance.

Applications
Oxyacetylene or TIG deposits used to surface rock cutting tools. Arc weld deposits used for sand mixer blades, pug mill knives, coal cutter picks.

GROUP 7 MATERIALS
Chromium boride paste
Chromium boride paste is spread over a steel parent material and is fused by carbon arc, MMA, TIG, MIG, or plasma welding.

Characteristics
Extremely high abrasion resistance.

Applications
Chutes.

Welding processes for surfacing

This section describes welding processes which are used for application of wear resistant coatings. Each has specific characteristics which are summarised and it shows that particular types of work can only be carried out by certain processes. However, the variety available does mean that some work can be done by several processes, which permits flexibility of choice. This can be important when undertaking repairs or replacements after a breakdown and in the field. This flexibility and the wide availability of process equipment and welding skills makes weld surfacing a popular technique.

PROCESS PRINCIPLES AND CHARACTERISTICS
Process principles and their characteristics are summarised for comparison purposes in Table 2.2, and are described in detail below.

Oxyacetylene
In this process flame adjustment is important and is described, Fig.2.1, by the length of the outer cone — the 'feather' containing excess acetylene — as

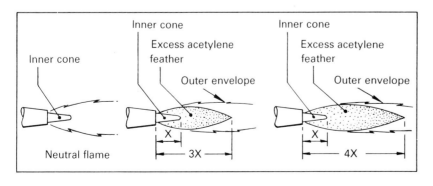

2.1 *Flame adjustment with length of outer cone (feather) as ratio of inner cone.*

Table 2.2 Weld surfacing processes

Process	Abbreviations	Approximate deposit thickness minimum, mm	Deposition rate, kg/hr	Dilution single layer, %	Typical uses
Oxyacetylene	OA	1.5	≤1	1-5	Small area deposits on light sections
Powder weld	PW	0.1	0.2-1		Small area deposits on light sections
Manual metal arc	MMA	3	1-4	15-30	Multilayers on heavier sections
Tungsten inert gas	TIG, GTAW	1.5	≤2	5-10	High quality low dilution work
Plasma transferred arc	PTA	2	≤10	2-10	High quality lowest dilution work
Metal inert gas	MIG, GMAW	2	3-6	10-30	Faster than MMA, no stub end loss; positional work possible
Flux-cored arc	FCAW	2	3-6	15-30	Similar to MIG. Mainly for iron base alloys for high abrasion resistance
Submerged-arc (wire)	SAW	3	10-30	15-30	Heavy section work; higher quality deposits than FCAW
Submerged-arc (strip)	SAW	4	10-40	10-25	Corrosion resistant cladding of large areas.
Submerged-arc (bulk welding)	SAW				Similar to SAW wire but other alloys possible
Electroslag (strip)	ESW	4	15-35	5-20	High quality deposits at higher deposit rates than SAW. Limited alloy range

Note: The spray fuse process is covered in Chapter 3

a ratio to the inner cone. The acetylene feather contains particles of carbon which tend to carburise the surface of the base metal and also reduce oxides on it.

Flame adjustments for the different types of surfacing alloys are:

1. Neutral; nickel base self-fluxing alloys.
2. Slightly reducing; carbon or alloy steels containing a high percentage of iron.
3. 1X-2X feather; deposition of Co base alloys on to austenitic stainless steels, to prevent carburising the base material.
4. 3X feather; Co base alloys on other substrates.
5. 4X feather; surfacing rods comprising tungsten carbides in a ferrous tube.

Advantages
1. Minimum melting of the parent metal is possible with low dilution of the surfacing alloy. This is advantageous when using highly alloyed consumables and is also important if a thin coating only is desired.
2. There is minimum solution of carbide granules from tubular rods.
3. The process is under close control by the operator.
4. Small areas can be surfaced.
5. Thin, smooth coatings can be deposited.
6. Grooves and other recesses can be filled accurately.
7. The operator can contour the deposit to minimise final machining.

Disadvantages
1. The process is slow and not suitable for surfacing large areas.
2. The build-up of heat may overheat the component and lead to distortion.
3. The range of iron base consumables is limited.
4. As it is usually a manual process, results are dependent on operator skill, fitness and degree of fatigue.

5 Good technique is essential to ensure that the bond at the interface is sound, especially when fusion with the substrate is not involved. There is a lack of NDT methods to check adhesion between coating and base material.

Powder welding

Powder welding uses a modified oxyacetylene torch fitted with an integral hopper to contain a self-fluxing surfacing alloy powder. The powder flow, under finger lever control, is entrained in the flow of combustion gases and emerges with the torch flame. The nickel base alloys normally used with this process can be deposited in thin layers and the technique lends itself to building up worn corners. Grit blasting to provide a mechanical key to the substrate is unnecessary as the deposit has a metallurgical bond to the substrate, but is often the easier method of ensuring that the surface is clean. Preheating the part speeds up deposition as the interface has to be brought up to the bonding temperature of 1000°C, without this the surfacing alloy is just cast on to the surface. There must be enough heat and time for the self-fluxing action of the alloy to clean the surface of interfering oxides to ensure a full metallurgical bond between it and the coating. Once the initial coating has been established the deposition of subsequent layers is fast and easy. By making sure that the initial layer is not too thick, deposits can be made on small areas of large components.

Self-fluxing alloys possess complex structures and have a wide temperature range between liquidus and solidus. In between these temperatures the alloy has a 'pasty' consistency and by careful control of heat input the deposit can be built up to thin, sharp edges, a good example being protection of the edges of cast iron moulds used in the manufacture of glass containers. Preheating the base metal to about 600°C, after applying a thin protective coating of the surfacing alloy on the cold metal, enables high deposition rates to be maintained and produces a deposit which needs minimum finish machining.

Advantages

This surfacing method requires less skill than a gas weld deposit and should be considered for surfacing small parts, small areas on large components or for generation of irregular deposits.

Disadvantages

1 There is a limited range of consumables suited to the process.

2 See under oxyacetylene welding.

Manual metal arc (MMA)

In manual metal arc welding, Fig.2.2, an electric arc is maintained between the electrode and the workpiece, the electrode consisting of a core wire (1.6-8mm diameter) with a flux covering.

The arc melts the parent metal and electrode to form a molten pool, which is protected from the atmosphere by liquid slag and gas formed by melting and vaporisation of the flux covering. The slag formed from the molten flux adheres to the weld surface, protecting it as it cools, and must be chipped away after each weld pass.

Advantages

1 It is adaptable to small or large complex parts.

2 Positional welding is possible, *i.e.* vertical.

3 It can be used with limited access.

4 A wide range of consumables is available.

5 Deposition rates up to 4 kg/hr are possible.

6 It is ideal for one off and small series work.

7 It is useful where only small quantities of hardfacing alloys are required.

Disadvantages

1 A skilled operator is required for high quality deposits.

2 Removal of slag is necessary, reducing duty cycle.

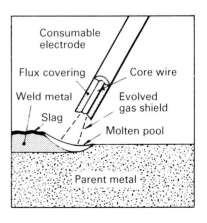

2.2 *Manual metal arc welding, also known as shielded metal arc, stick or electric arc welding.*

3 Dilution tends to be high.
4 Granular carbides in tubular electrodes are usually melted.

Tungsten inert gas (TIG)

In TIG welding an electric arc is maintained between the non-consumable tungsten electrode and the workpiece, Fig.2.3. The filler material is usually in rod or wire form.

The arc melts the parent metal to form the molten pool which is protected from the atmosphere by the inert gas shielding. Filler wire is fed into the pool manually to produce the surface coating.

Advantages
1 Low penetration is achieved with dilution 5-10% depending on technique.
2 The process can be closely controlled by the welder, and small areas can be surfaced, *e.g.* small pores in hardfacing deposits.
3 The process can be used manually with hand held torch and hand fed filler rod or mechanised for special applications.
4 A deposit thickness of 2mm upwards is achievable.
5 It provides a deposition rate of up to 2 kg/hr.
6 It is capable of high quality deposits.

Disadvantages
The process is not suitable for site welding and is generally restricted to a workshop free from draughts which affect the inert gas shroud.

Plasma arc

The plasma arc process, Fig.2.4, uses an argon shielded tungsten arc as the source of energy. A DC pilot arc is established between a central tungsten electrode within the torch, and a water cooled copper annulus which surrounds the electrode. The function of this pilot arc is to facilitate initiation of a heavier current transferred arc between the electrode and the workpiece when surfacing commences. A separate DC power source is usually connected between the tungsten electrode and the workpiece and controls the transferred arc which is constricted by a narrow orifice in the copper annulus.

This process, which is mechanised, is used to deposit alloy powder which is conveyed from a hopper to the torch by a carrier gas. It should not be confused with plasma spraying which uses a non-transferred arc to generate the heating plasma for spraying alloy or ceramic powders (see Chapter 3).

Advantages
1 It produces low penetration and dilution.
2 It is mechanised and provides close control of surface profile with minimum finishing required.

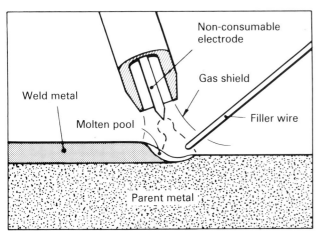

2.3 *TIG welding, also known as inert gas tungsten arc, tungsten arc gas shielded and gas tungsten arc welding.*

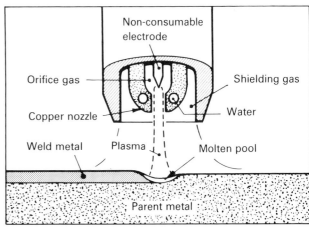

2.4 *Plasma welding.*

3 Deposition thickness is in the range 2-5mm.
4 The deposition rate is higher than with the TIG process, averaging 3.5 kg/hr.
5 The range of torches available includes those for surfacing bores down to 35mm diameter × 400mm deep and internal valve seats.
6 The process lends itself to sequence control of the weld cycle and also to automatic loading/unloading of workpieces. Thus it is an ideal production line tool.

Disadvantages
1 The equipment is not readily portable.
2 It is one of the higher equipment cost processes.

Metal inert gas/metal active gas (MIG/MAG)
In the MIG/MAG process an electric arc is maintained between the electrode wire and the workpiece, Fig.2.5. The parent metal and consumable wire are melted by the arc to form a molten pool, which is protected from the atmosphere by gas fed coaxially with the wire through the welding gun nozzle.

Advantages
1 It is a continuous process that is used semi-automatically with a hand held gun, or is wholly mechanised by traversing the gun and/or the workpiece.
2 Deposit thickness of 3mm upwards is achievable.
3 Deposition rate is 3-6 kg/hr.
4 No slag removal is required.
5 It provides a positional surfacing capability.
6 Controlled transfer pulse techniques produce less spatter and provide greater control of weld bead characteristics.
7 Guns are available for internal bore work.

Disadvantages
1 Use of shielding gun makes the process marginally less transportable than MMA and gas must be selected to suit the surfacing alloy.
2 Welding must generally be carried out within about 4m of the wire feeder, but best results are obtained with the gun as close as possible to the feeder, especially when feeding stiff wires.
3 Dilution may be high unless appropriate procedures are used.
4 High levels of UV radiation are produced especially when using high peak current pulse welding.

Flux-cored arc
Flux-cored arc welding is similar in principle to the MIG/GMAW process, but the core of the tubular electrode contains a flux which decomposes to provide a shield to protect the molten pool.

Advantages
1 It is a continuous process that is used semi-automatically with a hand held gun or is wholly mechanised.
2 A wide range of consumables is available.
3 No shielding gas is required.
4 Thickness of deposit is 3mm upwards.

Disadvantages
1 Dilution is 15-30% depending on technique, so it is not suitable for use with non-ferrous surfacing alloys.
2 Regular maintenance of equipment is necessary.
3 Deposit quality may be lower than the MIG/MAG process.

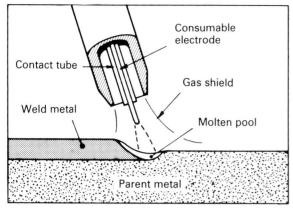

2.5 MIG/MAG, also known as CO_2, gas shielded metal arc or inert gas metal arc welding. Similar equipment is used for flux-cored arc welding in which a tubular electrode containing flux and alloy powders replaces solid filler metal.

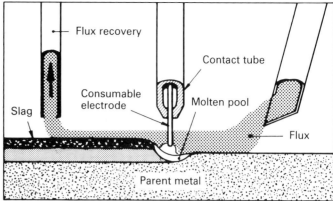

2.6 Submerged-arc welding.

Submerged-arc

In submerged-arc welding an electric arc is maintained between the electrode and workpiece, Fig.2.6. The arc, which is submerged beneath a covering of flux dispensed from a hopper, melts the electrode, the surface of the workpiece and some of the flux which protects the molten pool from oxidation.

The electrode is fed at a controlled rate to maintain a stable arc of constant length and flux which has not melted can be recovered and reused.

The process, which is mechanised, can use a wire or strip consumable and is capable of accepting alloy additions to the molten pool from a separate feeder.

Advantages
1 It is a mechanised process.
2 Deposit thickness of 3mm upwards is achieved.
3 Deposit rate is 10 kg/hr upwards on suitable workpieces.
4 A wide range of consumables is available.

Disadvantages
1 It is suited to large workpieces which can tolerate the high current and high deposit rate feature without overheating.
2 It is intended primarily for workshop use on a fixed installation.
3 Dilution may be high unless appropriate procedures are used.
4 Applications are generally limited to cylindrical or flat components; there is limited access to internal surfaces of larger bores.
5 Flux costs must be taken into account.

Wire consumables
A wide range of wires is available. Using careful overlapping of individual beads, large areas can be covered with smooth deposits which need little machining allowance.

Strip consumables
When covering large areas, the use of a strip consumable offers advantages in deposition rate. The range of materials available in this form is more limited than for wire, but the process is popular for applying corrosion resistant cladding to large vessels and process plant.

Alloy additions to the molten pool — bulkwelding
In bulkwelding, powdered alloy additions are metered into the weld zone and covered with flux. The consumable can be solid or tubular (cored) wire; the layer of powdered metal 'cushions' the arc and reduces substrate penetration and dilution. Precise control of the mixture of powdered alloy metal and electrode metal produces the required composition of weld metal.

One application of this process is production of wear plates consisting of high chromium iron deposits on a mild steel backing.

Electroslag
Electroslag welding uses equipment similar to the submerged-arc process for strip cladding. Deposition takes place under a blanket of flux, whose composition is formulated to produce a molten layer of slag by passage of the welding current. This, as well as protecting the metal transfer from the electrode to the substrate, provides sufficient heat to melt the electrode and to raise the surface of the substrate to a temperature at which bonding with the coating can take place. Materials available for use with this process are limited to those available in strip form, such as some stainless steels and nickel base alloys.

Compared with submerged-arc strip cladding, advantages claimed for electroslag include low defect levels, higher deposition rates and better control of dilution. The more popular use of this process is cladding large areas with corrosion resistant alloys.

Process performance
Table 2.2 shows typical figures for various process characteristics and it is important to bear in mind that results actually achieved depend to a large extent on the way the process is used, *i.e.* the welding procedure adopted. This subject is dealt with later in this chapter, but one characteristic of particular interest to the production engineer is the rate of deposition. Ranges quoted are typical of those used in practice and represent performance with a 100% 'arc on' time. Adjustment must be made for 'arc off' time in calculating an actual production rate. It is also important to recognise that the maximum deposition rate achievable on a given job depends on the extent to which dilution must be controlled, as this will increase as the component temperature increases; high deposition rates require high welding currents and these encourage a faster rise in temperature.

POWER SUPPLIES FOR ELECTRIC WELDING PROCESSES
Considerable development has taken place recently to improve the performance of power supplies, based on recognition of the need for accuracy and stable conditions to ensure repeatability of results to the desired quality standards. Conventional power supplies based on transformer/rectifier systems operating at mains frequency are now being replaced by units based on power electronics, capable of supplying waveforms suited to each welding process.

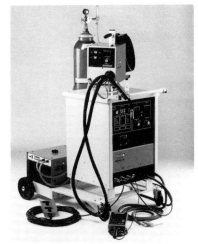

2.7 *GEC Control Arc 500 power supply for TIG, MIG/MAG/FCAW, MMA and air arc, and providing 30kW of power for pre- and post-heating.*

The new designs are compact and lighter than their predecessors and micro-electronic control enables output to be selected to suit each process so that the power supplies are truly multi-functional. The accurate control of pulse waveform necessary for controlled transfer pulse welding is provided by the new supplies which also allow the waveform to be varied automatically to a preset programme when a change in current is required, providing synergic control.

Examples of two such units are shown in Fig.2.7 and 2.8, both can be used for MIG/MAG, TIG, FCAW and MMA; the unit shown in Fig.2.7 is also capable of supplying up to 30kW of HF power for induction pre- or post-heating.

These power supplies are suitable for welding operations which need to be mechanised and can follow a predetermined weld cycle without operator involvement once parameters are established.

Metallurgical and related effects of weld surfacing on substrate and coating

This section deals with changes which may take place in the component material and alterations which can occur in the coating material as a result of the welding operation. The principal factor causing these changes is the temperature reached in component and coating during the process.

Most steels may have their properties changed by a controlled thermal treatment; temperatures used are similar to those reached during welding. Lack of control of the temperature cycle during welding may result in

2.8 *TPS Super 500 power supply, used with plug-in control units for synergic pulsed MIG, plasma, submerged-arc, TIG, MMA, FCAW, hyperbaric and air arc (courtesy Tom Palmer (Scotland) Ltd).*

hardening/embrittlement of the steel, with effects that may become evident during the process, or later in service. The problem can usually be overcome by controlling the temperature cycle so that the part remains in an unhardened and tough condition.

There are two main types of coating material; one is affected in its main properties by the temperature cycle like a hardenable steel, so the cycle must be controlled accordingly. The second type is not dependent on thermal treatment to acquire its properties, but in common with the first type, its reaction to the temperature cycle of welding may be affected by what is called 'dilution', *i.e.* intermixing of the deposit with the substrate material.

A further feature of weld surfacing is development of residual stress in the part. This arises from differences in thermal expansion characteristics between coating and substrate and from thermal gradients present during the coating operation. These stresses may result in distortion of the part or, in the case of the harder coating alloys, in contraction cracks in the coating.

A knowledge of the characteristics of the chosen substrate and coating materials is used to design thermal cycles for surface coating operations which produce acceptable finished parts.

THE TEMPERATURE CYCLE

Heat treatment before surfacing

Substrate materials should be specified in the soft condition for weld surfacing. The temperatures involved produce some tempering of the structure of a hardened steel. Parts made by casting or forging, or from bar stock that has been extensively machined should be stress relieved before the final preparation of the part, especially when distortion of the part during the welding process must be kept to a minimum.

The need for preheat

It is well known that molten metals shrink when they solidify. For castings, the pattern is typically made around 2% larger than the casting size needed.

Weld deposited metals shrink similarly unless — as is usual — they are restrained from doing so by attachment to the substrate material. This leads to residual stress in the component.

By applying preheat to the part, it expands and contraction on cooling after welding partially offsets weld shrinkage. Residual stress is thereby reduced and this helps to reduce any tendency to contraction cracking in a deposit of surfacing alloys near the top of the hardness range. It also helps to control distortion of the part by reducing temperature gradients.

In steel substrates designed to harden if cooled in air from their hardening temperature, a thin layer of the heat affected zone (HAZ) under the weld deposit may harden. This transformation involves a volume expansion which increases any residual welding stresses and can cause cracks both in the deposit and in the HAZ. The latter will cause detachment of the coating.

The use of preheat, maintained at a specific minimum temperature during the welding cycle, can ensure that no hardened HAZ is created during this period by reducing the speed at which the HAZ cools and by limiting its lower temperature while welding proceeds.

Depending on the transformation characteristics of the steel, this may be sufficient to allow the HAZ to transform to a soft structure, but a safe precaution is to arrange an immediate post-welding heat treatment which assures that only a soft structure exists throughout when cooled at room temperature. In oxyacetylene welding, somewhat higher temperatures may be desirable than the minimum required to avoid a hardened HAZ. This is because the flame is a low intensity heat source and by using a higher preheat, faster deposition is possible without risk of increased substrate melting and deposit dilution.

The upper limit of possible preheat is governed by several factors. The first is that the surface of the component may oxidise ahead of the weld, incurring the risk of defective deposits — porosity caused by oxide inclusions and/or improper bonding to the substrate. A limit is around 400°C for low alloy steels and 600°C for oxidation resistant materials.

The second factor is that with increased temperature, the strength of the substrate may be significantly less than the coating alloy and this may increase rather than reduce the distortion risk. A further risk is that of increased intermixing of the deposit with the substrate as the latter is more rapidly raised to its melting point by the heat source. The consequences of this effect — dilution — are discussed elsewhere in this chapter. High temperatures during welding can also lead to grain growth in some steels.

Interpass temperature control
Having stipulated the minimum and maximum temperatures at which weld surfacing can take place safely, as described above, the question of interpass temperature control is settled. In practical terms this may mean that deposition must be stopped periodically to avoid overshooting the upper limit or welding switched to a cooler portion of the workpiece, if this is possible.

At the other extreme, it means that at no time may any portion of the workpiece fall below the minimum set temperature, as the HAZ in that region may then transform and harden, with possible risks of cracking. This means that in practice, local preheat may be needed to provide adequate control on a large component. When depositing alloys designed to develop their hardness by a martensitic transformation, hardening may be prevented by the use of preheat or a rise in component temperature during welding. Specialist advice should be sought from the consumable manufacturers.

Post-weld heat treatment
For components made from hardenable steels, this can take two forms:
— Temperature equalisation on the welding set-up, followed by slow cooling in an insulating powder such as vermiculite.
— Temperature equalisation followed by immediate transfer to a heat treatment furnace preheated to the minimum interpass temperature. The subsequent temperature cycle can then follow various paths:
  i Cool to just below the martensite transformation temperature, hold, and then raise to temper the martensite;
  ii Raise to the austenitising temperature and then cool at a rate to produce a pearlitic (soft) structure;
  iii Raise to the temperature for most rapid transformation of any austenite to pearlite: hold for completion and then slow cool.

To select appropriate treatment times and temperatures requires a study of the available isothermal or continuous cooling transformation diagrams for the steel.

Subsequent heat treatment
Components that have been cooled in an insulating powder after welding are likely to contain more residual stress than those heat treated in a furnace. Heat treatment at temperatures recommended for stress relief of steels at around 650°C is effective in reducing movement when parts are subsequently machined and also for dimensional stability in use. Slow heating and slow cooling is needed for best results; cooling at 100°C per hour is often used.

Substrate materials — special precautions
Some steels may be embrittled by exposure to elevated temperatures; for example, 11-14% manganese steel must be kept cold by intermittent welding or water cooling. Some Ni-Cr steels can be susceptible to temper brittleness, or a reduction in the ductile-brittle transformation temperature, while some high alloy stainless steels can form the brittle sigma phase when welded. Specialist advice should be sought.

Other substrate materials
There is a wide range of engineering metals and alloys in use. When considering the possibility of applying a welded coating for improved service life, the first question to be asked is whether a cheaper substrate material of much superior welding suitability can be substituted — often at greatly reduced cost.

If the existing material must be considered, the following lists characteristics which may make weld surfacing difficult or even impossible:

— Melting point lower than surfacing alloy;

- Free cutting properties;
- Contains easily oxidised elements, *e.g.* Al, Cu, high Si;
- Is hard and/or brittle;
- A stainless steel not stabilised for welding;
- A steel containing nitrogen as a deliberate addition.

Grey cast irons
The melting point of cast iron is lower than that of steel and this limits the coating alloys that can be deposited without melting the substrate and diluting the deposit. Other problems include the risk of substrate cracking caused by welding stresses — it has a low tensile strength and, depending on composition, may develop hardened HAZs of even greater brittleness. Problems are increased by the presence of any casting defects. The oxyacetylene process offers the best possibility of controlling dilution, but use of a cast iron welding flux may be necessary.

Self-fluxing nickel base alloys (Table 2.1), Group 3, type 5 are capable of producing sound deposits of low dilution on good quality grey iron castings using powder welding. These alloys are nearer in melting point to cast iron than most surfacing alloys; in addition they possess a self-fluxing characteristic which promotes good wetting of the cast iron surface and a sound bond with minimum dilution. A widely used application is protection of the edges of tooling used in production of glass bottles, jars, *etc.*

White cast irons
White irons are extremely brittle and weld surfacing is not recommended.

Ni-hard cast irons
These irons are also difficult to weld surface, although some success has been reported by use of high levels of preheat, a buttering layer of a ductile material like nickel under the surfacing alloy and a very slow rate of cooling.

PRACTICAL RECOMMENDATIONS
Selection of thermal treatments for surfacing steel components must take into account the various factors mentioned above and in addition component size, shape and practical problems of handling the parts during and after welding, either through size or production volume.

INFLUENCE OF SUBSTRATE DEFECTS
The presence of defects in the material from which a component has been made is likely to affect the quality of the welded deposit. The importance of such defects depend on the quality standards demanded for the finished job.

Much surfacing work is carried out to improve the service life of parts used in critical items of equipment, including chemical and petroleum plant, power stations and heat engines. Standards for these applications are high and after finish machining/grinding parts are subjected to examination for freedom from defects in the coatings.

Two important sources of defects in welded coatings arising from the substrate are:

- Discontinuities in the substrate, *e.g.* porosity or cracks;
- Inclusions, *e.g.* mould material in castings, scale in forgings, slag particles/stringers, sulphur segregation.

Careful inspection of the substrate is therefore indicated, at a level appropriate to the quality of the finished part. Discontinuities are identified on the machined surface of the part before welding by dye penetrant inspection; defects should be excavated to verify that they are not an outcrop of a larger cavity. Inclusions should also be excavated to remove them entirely and the resultant holes filled by welding, using a consumable to match the substrate, and then dressed flush with the surrounding surface.

When defects become apparent during the welding process by local collapse of the surface of the component, attempts to 'weld them out' usually result in failure, either because of the appearance of porosity in the welded surface after machining, or even the presence of cracks caused by local increase in thickness of the coating material or to porosity acting as a stress raiser.

Repair of defects in welded deposits is dealt with later in this chapter.

DILUTION

Intermixing of the welded deposit with the substrate material — 'dilution' — results from the heat of the welding process melting the surface of the component material. The extent to which this occurs depends on the surfacing and substrate materials used, the welding process chosen and the parameters employed.

A feature of weld deposited coatings is the strong bond with the substrate and temperatures required to achieve this result in some melting of the substrate. The consequence of this intermixing is that properties of the surfacing material may be altered and its performance changed. Excessive dilution during surfacing is always to be avoided and the question arises as to how much can be tolerated and how it is to be controlled.

Many ferrous surfacing alloys supplied as electrodes or wires for arc surfacing processes, which are to be deposited on steel substrates, are identified by their typical weld metal composition when applied by specified processes and recommended welding parameters. As both substrate and coating are iron base alloys, the composition of the consumable can take into account some dilution and provide the desired weld composition and properties.

If the surfacing alloy is not iron base, but for example an alloy from Group 2 (nickel) or Group 3 (cobalt) base and the introduction of iron into the composition is known to affect wear resistant properties, the alloy designer must recommend that dilution is kept as low as possible for best results.

There are several ways in which this can be achieved:

— Use of a low energy heat source such as an oxyacetylene flame. By using the braze welding technique mentioned earlier with cobalt base alloys, it is possible to keep dilution below 5% on much work;
— Self-fluxing alloys containing silicon and boron are also deposited with similar levels of dilution by oxyacetylene.
— If a higher deposition rate is required, the PTAW (plasma transferred arc welding) process can be used and dilution levels similar to oxyacetylene are achievable on single layer deposits.

On components of variable cross section, or those whose temperature may rise significantly during welding (which encourages increased dilution), it is normal to regard figures of less than 10% dilution as realistic for many applications. This is more difficult to achieve with direct arc processes such as GMAW and SAW on single beads, but by use of overlapping beads, negative polarity and suitable parameters where appropriate, dilution of 15% or less can be achieved. If lower figures than this are required, a second layer deposit should be considered.

Apart from the influence of dilution of a surfacing alloy with iron from the substrate, other elements may be introduced from the same source. For example, an austenitic steel surfacing material dependent on a low carbon content for maximum corrosion resistance could be affected adversely by pick-up of carbon from the substrate.

Dilution between the coating alloy and substrate can sometimes produce undesirable properties. This is well illustrated in surfacing austenitic manganese steel with a material giving a hard surface as deposited. Interdiffusion causes the surfacing alloy to mix with the Mn steel reducing the Mn content so the substrate becomes brittle near the surface. A similar effect is produced when surfacing carbon and low alloy steels with an Mn steel. In this example the brittle layer is formed in the overlay. In either, the interdiffusion can be prevented by using a buffer or buttering layer of an austenitic stainless steel. Alternatively, hardfacing rods of a work hardening nature but containing 12%Cr and only 3%Mn can be deposited without formation of a brittle layer if reasonable precautions are taken. Other modifications can be used, the requirement being to preserve the austenitic structure even with some change in composition.

The figures for dilution given in Table 2.2 show a range because of the process variations mentioned above. In direct arc processes such as MIG and SAW, dilution is influenced *inter alia* by welding current, which also governs deposition rate. Hence these factors must be considered together when determining the process and parameters to be used.

COATING PROPERTIES AND STRUCTURE

It is important that the designer appreciates the variation in methods of deposition associated with each alloy group and the effect that this can have on the ultimate properties of the coating. Apart from dilution with the substrate material, both the micro- and macrostructure of the deposit can be changed by alteration in the surfacing method.

Surfacing rods containing tungsten carbide

Surfacing rods containing tungsten carbide produce different microstructures between arc weld and gas weld deposits. Higher heat input in arc weld depositing takes much more of the carbide into solution, hardening the matrix and reducing the amount and size of the carbide particles. The structure is also greatly influenced by the initial particle size of the carbide grains, Fig.2.9.

2.9 *Typical surface deposits obtained using different tungsten carbide grain sizes:* a) *Oxyacetylene deposit 60wt% WC, −10/+20 mesh, in 40% steel;* b) *Similar deposit using −80/+120 mesh carbide.*

Coarse particles give a better cutting action on rock, finer grades give better and more uniform resistance to wear. The filler metals comprise steel tubes filled with the tungsten carbide particles and are deposited with an oxyacetylene torch, or the usual arc processes. The melting steel takes up tungsten and carbon from the carbide to form a matrix anchoring the remainder of the carbide particles. The amount of carbide which dissolves in the steel depends on temperature and length of time the weld pool is molten. The extreme is reached when surfacing with very fine carbide using a high amperage electric arc. In this case all the carbide may dissolve giving a very hard brittle tungsten steel liable to weld cracking and containing few, if any, carbide particles. Much less solution occurs with gas welding and the carbide distribution remains more uniform.

With arc welding the surface of the substrate is melted and interdiffusion with the molten surfacing alloy provides the strong metallurgical bond. Fluxes may or may not be used, depending on the nature and amount of the oxide on the surface of the blank. Application temperatures are high, there can be much glare and fume from the flux adding to the difficulty of putting down an accurate deposit, so the designer must provide as much help as possible. Part of the surface melts so it is not feasible to design sections which are so thin they will melt through or taper to a feather edge.

Cobalt base alloys

Cobalt base alloys have lower melting points than steel and a different technique is used for oxyacetylene deposition on ferrous base materials. The flame is kept in a reducing state and the carbon from it alloys with the iron base substrate forming a skin of gradually reducing melting point until this reaches that of the surfacing alloy, around 1300°C, at which temperature the surfacing alloy can be melted on to the 'sweating' surface with which it forms a strong bond. Oxides created ahead of the pool float to the surface. The high chromium content of the surfacing material provides extremely good high temperature oxidation resistance so the deposit remains clearly visible and provided the blank design embodies good reference points the operator can build up the surface accurately. The alloy will develop reasonably smooth surfaces so the machining allowance is less than is required for arc weld deposition. On substrates which cannot produce a molten surface using a carburising flame a flux is necessary to clean off surface oxides to assist bonding to the surfacing alloy melted on to it.

A hypoeutectic Co base alloy (30%Cr-8%W-1.6%C) when deposited by gas welding shows dendrites of a Co-Cr solid solution infilled with a eutectic of

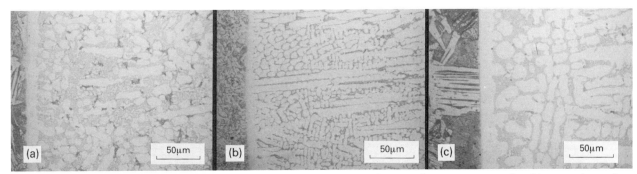

2.10 *Structures of cobalt base hardfacing alloy deposits:* a) *Oxyacetylene deposit of Group 3 type 2 alloy;* b) *Manual metal arc deposit of the same alloy;* c) *Oxyacetylene deposit of Group 3 type 1 alloy. For details of alloy groups see Introduction, Table 0.1.*

chromium carbides and the Co-Cr matrix, Fig.2.10. Deposited by arc welding the microstructure is finer and because of dilution by the base metal there is very little carbon to form carbides so there is little eutectic. The effect of a lower carbon content is seen more clearly in a Co base alloy (28%Cr-4.5%W-1%C), the percentage of the dendritic solid solution being much higher.

Nickel base surfacing alloys
Nickel base alloys vary greatly in composition. Ni-Mo-Fe alloys are used for corrosion resistance, Ni-Mo-Cr-W alloys for corrosion and heat resistance, both need high application temperatures to melt the alloy, whether gas or arc welding is used. The most widely used Ni base surfacing alloys contain boron and silicon which provide in-built self-fluxing properties. Melting at about 1050°C the molten alloy cleans off all but the most tenacious surface oxides from the substrate to which it bonds by a brazing mechanism, requiring negligible intersolution with the surface of the base metal. Flow of the deposit is controlled easily so the operator can cover complicated contours and produce thin smooth coatings on large or small areas. At the working temperature the deposition is easily seen, however, it is still essential that the blank design provides visual guides for position and depth of deposit. Formulation as well as deposition method may produce markedly different structures in Ni base alloys having the same chemical composition.

BUFFER LAYERS
The term buffer layer is used to describe the presence of an intermediate deposit between the base metal and the hardfacing weld material. The use of more than one type of hardfacing alloy may be necessary in some circumstances to reduce stress, to prevent cracking or improve wear life of heavy deposits. There are a number of applications where this practice occurs.

1 Hardfacing on soft material for high load service; when the 'harder' surfacing alloys are used on a soft base material, *e.g.* mild steel, there is a tendency for the hardfacing layer to sink in under high load conditions. Under extreme conditions this may result in the surfacing material spalling off. To overcome this a layer of strong, tough material is deposited on the workpiece under the hardfacing.

2 With gas welding techniques, if differences in thermal expansion are significant, and the surface hardness exceeds 50RC it may be useful to apply an alloy of compatible composition, hardness 25-30RC, underneath to prevent cracking in the hard overlayer. Where the design calls for a heavy build-up, full thickness may be achieved using alternate layers of hardfacing alloy and buffer material. In arc welding processes natural dilution from the base metal usually provides the necessary gradation of properties.

3 If a component is subject to heavy impact or flexing, there is a risk that deposits which do not relief check during welding will develop fine transverse cracks. These are not detrimental to the hardfacing but there is a danger that in service the cracks will act as stress concentrators and progress through into the base metal. This tendency is most pronounced when the base metal is a high strength steel. Use of a buffer layer prevents such crack propagation.

Designing for weld surfacing

Successful hardfacing depends on adequate surface preparation. Surfaces to be coated must be free from rust, grease, dirt or other material likely to be detrimental to the final deposit. Grinding, machining, filing, chipping, grit blasting, *etc*, are all used for preparation.

A sound base is required and in reclamation work this may necessitate removing fatigued or rolled over material, high ridges, or other major surface irregularities. Cracked or defective base material should be repaired. It is usually desirable to remove severely work hardened or case hardened surfaces before hardfacing.

For work, such as excavator digger teeth, bulldozer blades, *etc*, which are hardfaced by arc welding, extensive preparation is not necessary, as occasional blowholes in the deposit are not important. On precision components, such as engine valve seats, however, it is essential for the surfaces of the blank to be clean and the job should be smooth machined before welding. Rough machining marks are an indication of torn metal which may oxidise and cause pinholes in the deposit.

To prevent contamination of the deposit, sharp corners must be removed from the area of the blank to be hardfaced, as they are apt to melt if the arc or flame dwells upon them. This is particularly important if parts have to withstand corrosive conditions in service and which are being surfaced with iron free stainless alloys, or Ni or Co base alloys. Any serious contamination of the deposit with iron may lead to premature breakdown in service through corrosion and is likely to result in reduced hardness in the deposit as discussed earlier.

When building up edges of cutting tools, dies, *etc*, a recess is required to provide adequate support for the hardfacing material.

Wherever possible, the workpiece should be positioned so that hardfacing can be performed in the downhand position (workpiece horizontal). An uphill inclination of about 10° can sometimes be of assistance in laying down heavier weld passes. If work must be done out of position, detailed attention must be given to selecting suitable consumables and surfacing processes.

If facilities are available, blanks should be degreased immediately before hardfacing. Where surfacing is being applied as a reclamation or repair process, all cracks should be removed by machining, rough grinding or gouging and filled in using compatible electrodes, and machined smooth before final surfacing.

DEPOSIT THICKNESS

Weld deposits are generally at least 3mm in thickness and can be designed to any thickness required, although the harder, more brittle, alloys should be restricted to not more than two layers with a total deposit of about 8mm maximum to reduce the risk of contraction cracking. Oxyacetylene and TIG weld deposits can be thinner, down to 1mm thickness. The advantage of thick coatings is that they provide longer service life before resurfacing is necessary, assuming that the change in dimensions as the part wears can be tolerated. It is bad economics to design a deposit thicker than the allowable wear. Although components such as shear blades, gear wheels, conveyor screws and extrusion dies show a reduction in efficiency as wear takes place, other components such as gyratory crushers can tolerate wear of many centimetres before resurfacing becomes necessary.

If the deposit is to be ground it should be at least 1mm thickness and in most cases preferably not thinner than 1.5mm. During deposition some distortion of the blank probably occurs and also scaling of the centreholes or chucking or locating faces. Consequently, it may be difficult to ensure that the deposit cleans up to an even thickness all over. It could be ground off entirely in some areas unless it is more than 1mm thickness. On larger jobs, such as dies 100mm diameter, even a 1.5mm deposit (after grinding) is not sufficient to take care of possible variation in distortion or contraction and 2.5 or 3mm of finished deposit is usual.

In addition to the need to design a deposit thick enough to allow distortion to be corrected by machining and still leave the desired thickness for the working life of the part, it must be remembered that welding generally produces an uneven surface profile which may measure up to 2mm peak to trough in a manual weld and less with mechanised surfacing.

If the deposit is heated in service, such as on hot shear blades, it must be designed sufficiently thick (usually 3mm minimum) to ensure that the base metal is not so heated that it becomes soft enough to allow collapse of the deposit. Again, a thicker deposit is needed on areas where high pressures are exerted, such as the line of entry or throat of a drawing die.

Examples already quoted show how design calls for variation in the thickness of deposit depending on the amount of wear predicted in the various sections of a part or assembly. For reasons of economy the thickness of the deposit, after any necessary finishing, should be the minimum required for adequate component performance. The optimum thickness varies greatly from one application to another, since the 'wear tolerance' for the part may vary from as little as 0.05mm for a precision component to as much as 50mm for earthmoving equipment.

Process considerations can limit the minimum thickness which can be deposited. Guidelines for the designer are given in Table 2.2.

Finally, it should be borne in mind that when designing coatings for wear resistance the finished coating thickness should be greater than the permitted wear tolerance. The coating should never be allowed to wear through so that the base metal becomes exposed, or further wear will be very rapid. In detailing the extra thickness the designer must make allowance for the variation in deposit thickness which may occur across the part because of distortion in processing, as already discussed.

BLANK PREPARATION AND DEPOSIT DESIGN

General

Reference has already been made to effects of welding temperatures on steel components, *i.e.* some oxidation of adjacent surfaces and possibly some distortion, depending on the job. So for components which require finishing to accurate limits with a clean machined/ground surface, a machining allowance is necessary on all unwelded surfaces.

If a deposit is to be machined to a sharp corner, difficult corner build-ups by welding can be avoided by making the part locally oversize, later machining back to size and to produce the corner. If a weld deposited surface is to be precision finished flush with adjacent surfaces, then a recess should be provided for the deposit. This has the added advantage of providing a ready guide — especially to a manual welder — to the exact location and thickness of deposit required. Recesses should never have square corners, since protruding corners heat more rapidly than surrounding surfaces and tend to melt, diluting the deposit and reducing hardness, wear resistance and corrosion resistance of many surfacing alloys. Recessed sharp corners are slow to reach welding temperature even when the adjacent surfaces have done so and can be the site of defects such as porosity and an imperfect bond to the substrate, which may be exposed on machining.

When designing blanks, always provide a reference for the machinist to pick up, such as centres, shoulders or faces. Once the part has been welded, the marking or recess provided for the welder is obscured and with no reference, the deposit may easily be machined so far that it is unable to do the job; indeed some deposits have been totally removed because of this lack of foresight. Remember, the accuracy of the finished job depends on a correctly designed blank, Fig.2.11.

Design details

For welded coatings on flat or circular sections, the recess, Fig.2.12, should not be less than 13mm wide even for a shallow gas welded deposit on a thin section blank. The rim of metal at the side should be at least 3mm even when using a small flame. The depth should be the finished thickness plus grinding allowance. Any surface to be coated should preferably be smooth machined.

To prevent contamination of the deposit, sharp corners must be removed from the area of the blank to be hardfaced, as they are apt to melt if the flame

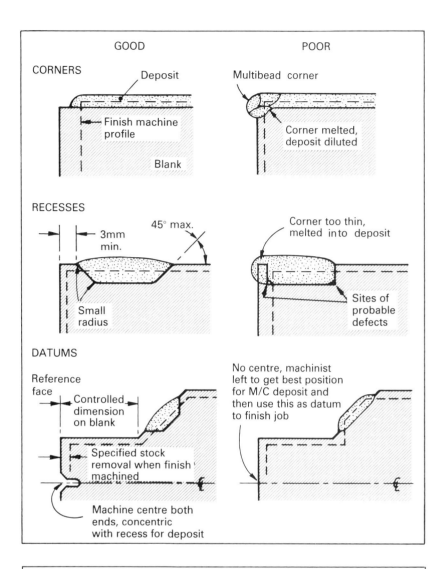

2.11 *General principles of blank preparation.*

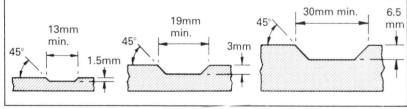

2.12 *Design details for welded coatings on flat or circular sections.*

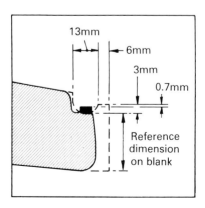

2.13 *Steam throttle valve — dotted lines show preparation of blank.*

dwells on them. This is of particular importance on jobs which have to withstand corrosion and which are being surfaced with iron free stainless alloys such as Co base materials. Serious contamination with iron could lead to premature breakdown caused by corrosion.

Figure 2.12 indicates minimum recommendations for various thicknesses of deposit. A 45° angle should be used for the side of the recess so that there is no corner to melt easily. If a radius is used it should be at least 3mm. The rim of metal must not be less than 1.5mm even on a small job needing a small flame and with a large flame 6mm may be required to prevent melting.

If a large component is being manually surfaced by welding it is locally red hot and the operator will have difficulty in deciding the exact location and extent of the area to be surfaced. A recess provides desirable assistance helping to control thickness of the deposit, preventing waste of alloy and reducing subsequent machining time and cost. A typical example for gas welding is a steam throttle valve, Fig.2.13. This illustrates several design features:

1   The recess for the deposit is radiused smoothly;

2   For oxyacetylene welding the recess is wide enough to take the large welding flame necessary to hardface the mass of metal. With a narrow

recess the two top corners would probably be melted before the correct temperature was attained in the base of the groove;

3 The outside wall, which is machined off after facing, is thick enough to prevent its melting during deposition;

4 The top of the recess indicates the correct height of the deposit. A smooth, level surface can be put down keeping excess surfacing material and finish machining to a minimum.

Small diameter components, Fig.2.14, can have an oversize deposit of metal laid on the end (*a*) if the sharp corner is chamfered to prevent its melting. Alternatively if the protection does not need to cover all the end a shallow depression can be formed (*b*). A groove with thin walls should not be designed into the part (*c*). On larger diameters a machined recess can be used (*d*) but there must be adequate metal left on either side. An alternative is to use a shallow groove (*e*) which avoids the side wall and so can use a less massive blank with savings in steel and finish machining.

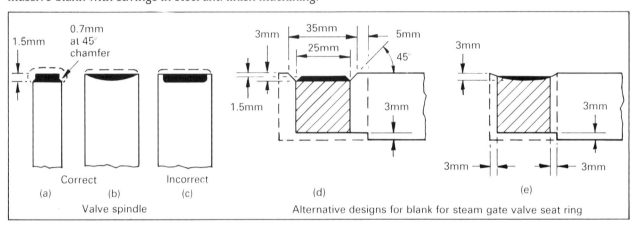

2.14 *Design of blanks for surfacing:* a)-c) *Smaller diameters — a and b correct, c incorrect;* d) *Larger diameter, machined recess;* e) *Larger diameter, shallow groove.*

Rock crusher hammers, Fig.2.15, are hardfaced to improve life. Wear on these starts at the leading corner, requiring the thickest coating there, tapering off down the front face. It should be borne in mind that the life of the hammer will be increased a number of times by corner protection so the extent of surface protection should be designed accordingly. The end face of the hammer need not be hardfaced; it does little work and when the front corner wears through it would probably be stripped off. A thin deposit is, however, sometimes of advantage down the side faces if they wear rapidly. Surfacing, as in Fig.2.15*a*, could cause premature failure of the coating on the other corner.

Distortion control
Contraction of the weld deposit during solidification and cooling can cause distortion of the component. This problem is more common in weld surfacing than in fabrication welding because there is little chance of balancing welding stresses by welding both sides of a neutral axis. There are three main factors:

1 Contraction of the weld metal during solidification and cooling from a temperature which may be considerably above that of the substrate. Shrinkage is approximately 2% on cooling to room temperature and this tends to pull the workpiece in an arc along the direction of the weld run, Fig.2.16.

2 Different degrees of expansion and contraction between the metal adjacent to and at a distance from the weld run. This problem is accentuated most on thin base materials. The metal close to the weld zone becomes hot and starts to expand. Being restrained by the colder stronger metal further from the weld zone it can only move by buckling around the weld area. This movement will not be fully reversed during the cooling cycle and the workpiece may remain permanently distorted. Alternatively the hotter metal may undergo plastic deformation and on cooling will not stretch again and so cause distortion, such as dishing of a valve seat.

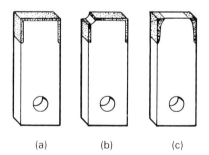

2.15 *Hardfacing of rock crusher hammers as in* a) *will break and fail rapidly, as in* b) *where the top coating has been stripped off;* c) *For economy the hammer is protected on both edges so that when the first deposit wears, the hammer can be reversed.*

2.16 *Shrinkage of weld metal during solidification and cooling tends to pull the workpiece in an arc along the direction of the weld run.*

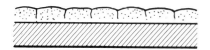

2.17 *Small 'relief check' cracks which form across the weld bead reduce the amount of stress distorting the base metal — often such cracking is not detrimental to the performance of the component surface.*

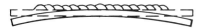

2.18 *To combat distortion a workpiece can sometimes be preset or prebent in the opposite direction.*

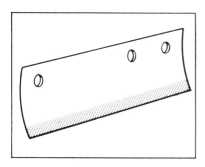

2.19 *For surfacing grader blades or bulldozer cutting edges place two blades back to back and prebend by clamping at each end with a spacer bar in the centre. Skip weld, working on each blade alternately applying a deposit 37mm wide in one run.*

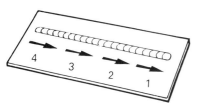

2.20 *Backstepping can reduce distortion when hardfacing a long blade.*

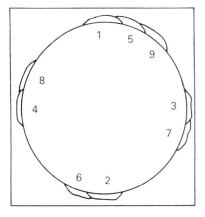

2.21 *To prevent warping when hardfacing shafts, a short run should be made parallel to the axis and then a similar run made on the opposite side to equalise stresses.*

3 Apart from temperature differences there will be differences in the coefficient of expansion and contraction of the surfacing alloy and the substrate. With many combinations this is relatively slight, but it can be an important factor with Ni and Co base alloys, cast irons and materials containing an appreciable percentage of tungsten carbides.

Often distortion, especially of the first type considered above, can be overcome by restraining the part so that it is not free to move. This can be done by clamping or tack welding the part to a firm support. For flat items, such as crusher jaws, two parts can be clamped or tack welded back to back and hardfacing applied to each alternately.

Surfacing using hard alloys may lead to cracking in the deposit if such restraint is used, Fig.2.17. If such cracking cannot be tolerated, a high preheat may prevent cracking in heavily restrained parts. Temperature may have to be up to 400°C for arc welding and 650°C for gas weld surfacing. The use of a soft buttering layer may reduce stresses between the overlay and the substrate. Multilayer deposits of hard alloys are sometimes made with alternate layers of a soft ductile metal to prevent propagation of cracks, a corrosion resistant interlayer may also eliminate corrosion of the substrate if this would occur through the cracks in a hard wear resistant top coating.

By slowing down the cooling rate and thus the rate at which hot metal contracts, preheating can sometimes be used to prevent, or certainly reduce, distortion. This allows more time for stresses to become equalised in areas immediately adjacent to the weld rather than distorting the workpiece.

Presetting or prebending the part in the opposite direction to that in which it would distort can be used effectively on metal up to about 12mm thickness, Fig.2.18. Contraction stresses pull the surface straight. The amount and actual nature of the presetting required for a given job must be established by experience. An alternative to presetting the part, illustrated by the surfacing of bulldozer cutting edges, Fig.2.19, and knife blades is to clamp the two components back to back with a spacer bar in the centre. Contraction of the surfacing alloy on release pulls the blade straight and puts the deposit into compression, a desirable state. A single blade could be clamped to a strongback with a spacer under its centre to make it convex during hardfacing.

The use of intermittent welding techniques, such as backstepping may greatly reduce distortion when hardfacing a long blade, Fig.2.20. Several centimetres at the centre should be hardfaced first and similar lengths on each side of this deposit should be laid down. The direction of deposition should always be towards the centre. Each run should be of the maximum width and thickness possible or necessary, so as to reduce to a minimum the number of times the blank is heated and cooled. Again if in surfacing a pair of components they are fixed back to back, distortion is minimised if hardfacing is performed alternately on each.

Similar considerations apply to manual hardfacing of shafts, Fig.2.21. To prevent warping a short run should be made on one side, parallel to the axis, then the shaft should be turned over and a similar run made on the opposite side to equalise the stresses. If a long run were made on one side only, this would cause a permanent set which would not be removed by the next deposit on the opposite side. The sequence of runs should be as shown in Fig.2.21. If the length of the area to be hardfaced is over 200mm long, it should be deposited in 100 or 150mm sections in the above manner. The job must be well supported at close intervals to prevent sagging, particularly if it is preheated to a red heat.

The problem of distortion of shafts is greatly reduced by mechanised deposition. The shaft is rotated about a horizontal axis and a continuous overlapping bead of weld applied in a controlled spiral. Mechanisation, when possible, usually helps to reduce distortion as a result of more uniform heat input and deposit thickness than is possible manually; economics in machining allowance are therefore possible.

The allowance which must be made for the effects of distortion, especially if the part has to be machined to accurate dimensions, depends on many factors, and must rely greatly on experience.

If a plate is preset it is better to overdo the convexity so that on flattening the deposit is compressed. It must be remembered that if straightening is carried out hot, perhaps at red heat, further distortion occurs on cooling to room temperature, so the process should leave the blank still slightly convex, to the order of 0.25mm per 100mm. Shrinkage, as well as producing bending in a plate, causes some shrinkage in the base material. The designer should make the blank overlong so that, where necessary, it can be machined back to size. As a guide there should be a minimum of 1% added to the length.

When hardfacing the bore of a die or the periphery of a sleeve, radial shrinkage takes place. The amount of shrinkage depends on the thickness and extent of the deposit. Other factors are the compositions of blank and surfacing alloy, the amount of preheat used and the temperature to which the blank rises. All materials lose strength as they are heated, some to a much greater extent than others, and the size of the blank has a great effect on the resistance it shows to deformation. Thus with gas welding where the heat input is greater and more widespread than with arc welding, distortion caused by plastic deformation of the blank is likely to be considerably greater. An austenitic steel is much stronger at elevated temperatures than a carbon steel so will distort far less. A thicker blank shows less contraction especially if the temperature is kept low during deposition so that the colder, stronger sections can support the hotter, weaker areas. A narrow band of surfacing on a long sleeve provides less contraction both on diameter and length than a much wider deposit. A figure of 1-2% is a reasonable first estimate.

Many components are so shaped that correction of weld shrinkage distortion is not possible by application of force, as mentioned in the example of a plate or shear blade. In such cases, it is necessary to envisage the effect that welding stresses will have on the shape of the part and to premachine it to a shape that will be largely self-correcting. One such example is shown in Fig.2.22 which depicts a ring that requires a deposit on one face to form a valve seat or sealing ring. Weld shrinkage will cause the ring to contract in diameter and at the same time, shrinkage on the deposited face will be greater, leading to the shape shown. By premachining an angle on the top face before welding, the deposit steel/interface becomes nearly flat and the risk of a locally thin deposit is avoided. An element of trial and error is involved in deciding the actual shape of a given component

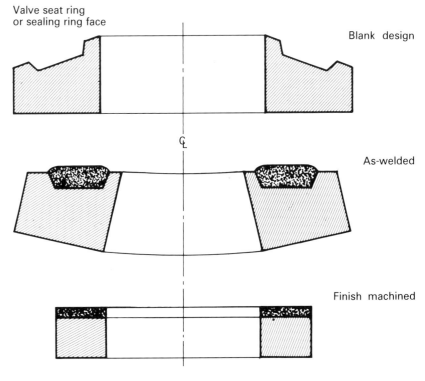

Note: proportions exaggerated for emphasis

2.22 *Blank design to accommodate welding distortion on rings.*

before welding, depending as it does on the finished size, shape and material used, as well as the process and weld parameters employed.

A second example, Fig.2.23, is that of a hollow cylinder to be faced on the inside diameter. Shrinkage will tend to reduce the diameter of the part but this will not be uniform along the length. Typically, the ends will shrink more than the middle, so that a taper premachined in the ends of the bore provides compensation and the finished deposit is substantially uniform in thickness.

Internal deposit in bore

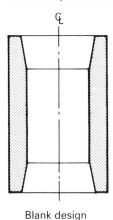
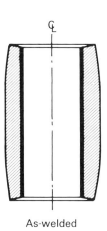
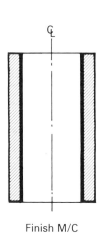

Blank design   As-welded   Finish M/C

2.23 *Blank design to accommodate welding distortion in bushes.*

In all these examples, the result will be more uniform and predictable when the job is mechanised; machining allowances on both the component and the welded deposit can be reliably reduced with savings in materials and time.

WELD PATTERNS
The actual pattern used in the welding operation can greatly affect the efficiency with which the surfaced component resists wear. This is an aspect of design which requires considerable experience and it cannot be reduced easily to design rules. A few example, however, may give an indication of some of the factors involved.

Hardfacing deposits are generally applied in one of three patterns. These are continuous cover, stringer beads or individual dots. Selection of the pattern to use depends on a number of factors including function of the component and service conditions.

Continuous coverage is used for hardfacing parts which have critical size or shape, such as rolls, shafts, tracks, crusher jaws and cones. It is often required on parts subject to much fine abrasion or erosion. Typical examples would be pump and fan impellers, sand chutes, valve seats, dredge bucket lips and pug mill augers.

2.24 *For protection against fire abrasion weld runs should be at right angles to the direction of travel.*

For protection against fine abrasion or erosion, weld runs should be at right angles to the direction of travel of the abrasive material, Fig.2.24, and care should be taken that sufficient overlapping of weld runs is given to ensure adequate coverage of the area being treated, Fig.2.25.

This design feature does not apply if a very hard deposit is being laid down and relief cracking is intended to eliminate excessive, harmful locked-in stress in the deposit. Electrode movement should be parallel to the direction of flow of the fine abradent as the cracking will be at right angles to this and across the direction of flow.

Stringer beads are often used when it is not necessary to cover the base material completely. Typical examples are dragline buckets and teeth, ripper teeth, rock chutes, *etc*.

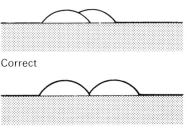

2.25 *There should be sufficient overlapping of weld runs if protection is required against fire abrasion.*

There are a number of fundamental guidelines for the designer. If surfacing of shovel teeth, Fig.2.26, is considered hardfacing one face will not be effective, since, as the unprotected edge wears away, the hardfacing will tend to chip off because of lack of support, Fig.2.26a, however, coating both faces, Fig.2.26b, will not give optimum results either because as the hard surface wears cavitation of the substrate is probable with the likelihood of chipping

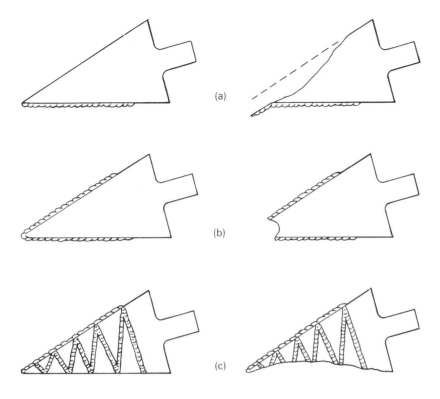

2.26 Choosing weld bead patterns for maximum life in service: a) Single surface, unsatisfactory; b) Both surfaces, also unsatisfactory; c) Optimum solution.

of both deposits. The best results are found with a surfacing pattern shown in Fig.2.26c where additional protection of the sides greatly assists resistance to wear and hardfacing the top, which is the more highly stressed by the abradent, reduces the rate of abrasion of the base material and is a self-sharpening action with the gradual wear along the bottom.

An example of the economical use of deposit patterns is illustrated in Fig.2.27. This shows a feed screw used in fertilizer manufacture handling granules containing limestone. The screw is 2.1m (7ft) long and 0.8m (2.5ft) in diameter and the face of the flights are deposited in a lattice pattern with a chrome carbide applied by manual metal arc.

For teeth working in coarse rocky conditions, it is desirable, Fig.2.28, for the stringer beads to run parallel to the path of the material being handled. This allows the large lumps of rock, *etc*, to ride along the top of the hardfacing beads without coming into contact with the base metal. For teeth working in fine sandy conditions the stringer beads are better placed at right angles to the direction of travel. This allows the fine material to compact in the intermediate spaces and provide protection to the base metal. Usually, however, a mixture of coarse and fine material is encountered and for these conditions a combination pattern, known as 'checker' or 'waffle' is used. This cross-hatched pattern is particularly effective where the material being handled is damp or stiff, *e.g.* clay, but it can be ineffective where the material is dry or free flowing. Performance achieved depends greatly on the spacing and height of the weld beads and should be determined experimentally for each application.

For areas where wear is less severe, such as along the rear of buckets, shovels, *etc*, a dot pattern, Fig.2.29, is often used. This reduces cost of material and application time. While not as effective as a waffle pattern it may allow material to compact between the dots and the high spots offer protection against abrasion from large lumps of rocks.

Generally the dots are 15-20mm diameter × 10mm high at about 50mm centres for earthmoving applications but there can be great variation. One specification for quarrying, also for damp material, using material containing tungsten carbide, is 20mm high with the dots spaced more closely, 6-30mm apart. Hard and fast rules cannot be formulated.

A dot pattern is useful on manganese steel castings to keep down heat input, and also to restrict excessive heating of the substrate when surfacing quenched and tempered steel.

2.27 *The life of an unprotected carbon steel feed screw, used in fertilizer manufacture, is about one month. A chrome carbide deposit in conjunction with an MMA deposit of tungsten carbide on the crest of the flights increases life to 9-12 months (courtesy ICI plc).*

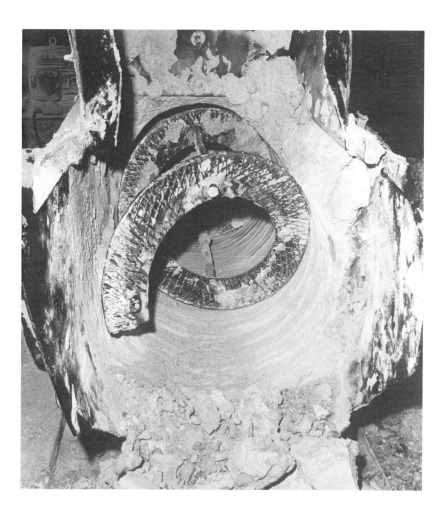

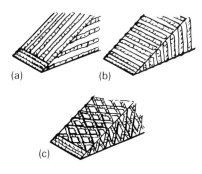

2.28 *Pattern of hardfacing for shovel teeth used in handling coarse rock: a) Teeth working in coarse, rocky conditions; b) Teeth working in fine, sandy conditions; c) Teeth working in coarse and fine material.*

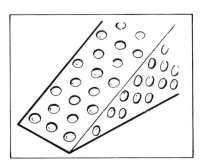

2.29 *A dot pattern can be used for areas where wear is less severe.*

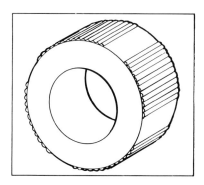

2.30 *Crusher roll with transverse beads applied.*

In the example of the crusher roller shown in Fig.2.30, transverse beads 12.5mm wide are applied. Skip welding is used to avoid overheating. The use of a positioner permits welding to be carried out much more easily. Longitudinal beads assist in gripping the material, but such a pattern would be most undesirable in, say, a pump where turbulence would result in decreased performance. For this, complete coverage should be used with the beads deposited parallel to the direction of flow.

For the most severe wear conditions complete hardfacing is needed and this may be supplemented by superimposed dots of alloys containing chromium carbide or tungsten carbide. This pattern is useful where hardfacing is subject to high tensile stresses, e.g. on rolls, as it can avoid notches which can introduce the risk of fracture.

Where the edge and face of a component is built up with a multiplicity of weld beads, Fig.2.31, it is essential to specify an adequate deposit on the corner otherwise rapid wear can take place at the edge.

MACHINING ALLOWANCES
Part of the design process involves decisions on machining allowances and these depend on two factors:

- Any change in shape of the component itself caused by welding stresses;
- Variations in deposit thickness, normally greater on manual deposits than those that are mechanised.

The tendency for distortion and shrinkage of the substrate material and the importance of providing datum faces for subsequent machining operations have been dealt with earlier in this chapter. Because of the variety of circumstances that are met in practice, no design figures can be quoted that will apply to actual jobs, and experience with a given type of work has a key role in blank design.

For work likely to be repeated, the following approach is suggested.

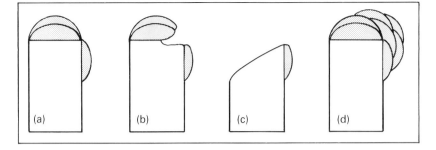

2.31 *Use of several weld beads to build up the edge and face of a component:* a) *Inadequate deposit on the corner;* b) *Resultant rapid wear;* c) *Undermined top surface layer breaks off;* d) *Correct method with extra beads deposited along the edge.*

1 Apply reasonably tight dimensional limits to the blank dimensions, *i.e.* minimise variations in size between blanks;

2 Make the surfacing operation as reproducible as possible, *e.g.* by mechanisation and use of stipulated and controllable welding parameters;

3 Ensure that post-welding heat treatments are always reproduced accurately;

4 Changes in shape and size of the job are measured and recorded and used to make controlled adjustments to the blank design of subsequent parts; the object being to economise on substrate and surfacing material, welding and machining time.

When parts having a low chance of repeatability are involved it is likely to be cheaper to err on the generous side of allowances, as additional cost is likely to be less than the cost of rectification if the part fails to meet drawing requirements after the first attempt.

It usually pays to take a proof cut over the deposit before machining unsurfaced areas. If a need arises to resurface locally, because of a low spot or area, there is then a finishing allowance still on the unsurfaced areas to cope with any small additional distortion from the second welding operation.

When choosing datum faces for the operator to use when machining the coating, choose one as close as possible to the deposit, to minimise errors which may result from shrinkage or distortion of the welded part.

COATING AREA
It is important to define clearly the area of the component which is to be treated. It must not be too small, as serious wear may occur outside the surfaced portion; but it should not be unduly generous as this adds unnecessary cost. A detailed knowledge of the operation of the component is often essential to obtain optimum results as actual positioning of the coating can greatly affect the efficiency with which the part performs. This aspect of design cannot be reduced to simple rules as it depends on knowledge of the individual application or examination of the worn part.

When a coating is designed into a wear-vulnerable area and successfully reduces wear at this location, it may be found that other areas of the component — which before had not given wear problems — may start to wear to such an extent that their life determines the life of the whole. This was illustrated dramatically with a sand muller. The machine incorporated a carbon steel shaft mounted vertically in upper and lower bushes. Installed without any hardfacing, it was found that after 280hr operation the shaft was running eccentrically as the upper bushing and rubbing area were badly worn. The lower bushing and shaft, however, had worn very little because they were enclosed and well protected from sand, grit and other abrasive materials. The upper bushing had considerably less protection. Because wear was slight in the lower section it was decided to hardsurface the worn area only. This involved a hard Ni base alloy applied by spraying and fusing.

After nearly 2000hr the shaft was again found to be running eccentrically. This time it was found that the lower bushing and shaft had worn. Little wear had taken place on the upper bearing area. Surfacing the lower section of the shaft ensured many thousands of hours of satisfactory operation.

Conveyor screws, augers, extrusion screws, *etc*, provide varying wear patterns in different sections along the length. Pressures rise at the exit end requiring harder coating materials, thicker deposits and more extensive coverage of the face of the flight in many circumstances. The central shaft may also need protection, possibly from abrasive wear or deterioration which is caused primarily by corrosion. The extent to which different areas need protection is often determined by experience but the designer needs to be aware of the probability of wear and take as many precautions as possible.

PHYSICAL AND MECHANICAL PROPERTIES OF COATINGS
The manufacturers of surface coating materials for application by weld surfacing publish data about their products. These figures may represent the undiluted weld metal or the deposited metal applied according to the maker's instructions.

Generally, products are designed to provide resistance to various forms of wear and/or corrosion, possibly over a wide temperature range. To achieve these properties, hardness, structure and corrosion resistance take precedence over tensile strength, elongation and other mechanical characteristics. It is normally assumed that the latter properties are provided by the material from which the body of the part is made and that this will provide adequate support for the coating.

Coating materials designed for resistance to wear usually feature hardness as an important characteristic and sometimes, the strength and toughness of these diminishes as the hardness increases. However, this seldom gives a problem in service, although a part exposed to a significant degree of fatigue loading might develop cracks in the coating during use. A softer, tougher coating material may be a compromise solution.

The question of dilution of deposit with substrate material has already been referred to and this is a variable effect depending on welding process and parameters used. It can be expected to modify the behaviour of the coating alloy from properties quoted for undiluted deposits. With cobalt and nickel base alloys deposited on low alloy steels, the pick-up of iron causes a drop in hardness, an increase in toughness, a drop in erosion resistance and some change in corrosion resistance depending on the corrosive medium involved. These changes are relatively small at dilution levels in the region of 5%, a level commonly achieved in good quality deposits. For many purposes, up to 8-10% dilution can prove acceptable in service.

Compared with cobalt and nickel base alloys which contain small amounts of iron, many iron based surfacing consumables have compositions which allow for the dilution associated with the welding process, *e.g.* MMA electrodes. In coating alloys like austenitic steels, dilution can affect corrosion resistance because of a reduction in the effective chromium content or an increase in carbon content through pick-up of carbon from the steel substrate.

Reference has been made earlier to the influence of welding conditions and process on the metallographic structure of deposits. Coating alloys which show large primary carbide development when applied by gas welding (a slow cooling rate) may have this suppressed by the faster cooling rates of some arc processes and this could lead to an increase in the toughness of the alloy deposit, and a reduced risk of contraction cracking.

Practical considerations
MANUAL VERSUS MECHANISED SURFACING
Weld surfacing originated from a need by plant users to restore worn parts to a usable condition and if possible to extend life between breakdowns. For this purpose, manual processes may be appropriate to the infrequent nature of the work; however some variation in quality of the deposits may be a consequence of this approach. The creation of a demand for parts bought as replacements to have deposits of wear resistant alloys incorporated in their manufacture naturally led to the need for mechanised surfacing processes to deal with the larger batch quantities as efficiently as possible.

Important benefits of mechanised processes are as follows:
— Higher rates of deposition, greater productivity;
— Greater precision of weld bead shape and position;
— Less surfacing material used;

- Smaller machining allowances needed, and possible saving on machining time;
- Improved product uniformity and performance;
- Fewer rejects and rework;
- Less operator skill needed;
- Reduced operator fatigue and discomfort.

Certain weld surfacing processes are unsuited to manual operation, such as submerged-arc, which is widely used for surfacing new parts and for repair work. Mechanisation of some manual processes can be carried out but it is those processes which use a continuous form of surfacing consumable which lend themselves best to mechanisation. The manual metal arc (MMA) process is not a candidate for mechanisation and the oxyacetylene process, although capable of mechanisation, is being displaced by electric arc processes.

Apart from submerged-arc referred to above, those most suitable for mechanisation are TIG/GTAW (Fig.2.32), PTA, MIG/GMAW and FCAW (see Table 2.2). The first two are non-consumable electrode processes and require separate adjustment of welding current and consumable feed rate. The other two regulate consumable feed rate in accordance with welding current. A consequence of this difference between the two groups is that the latter will automatically compensate for small changes in gun to workpiece distances while the former may require a separate arc length control mechanism.

The principal items of additional equipment required to mechanise a process are those which provide relative movement between torch and workpiece.

A positioner is used to rotate a circular or cylindrical workpiece at the required surface speed. The example shown in Fig.2.33 is fitted with a tilting table and variable speed drive. Such a unit is not required if, for example, weld deposits are to be applied only to flat surfaces.

2.32 A TIG coating of a cobalt base alloy of Group 3 type 2 on a guide roll showing the smooth controlled profile of a mechanised weld deposit.

2.33 Column and boom manipulator and tilting table positioner for surfacing rings and bushes (courtesy F Bode & Son Ltd).

A manipulator is used to control the position of the welding head and to provide movement to this during the welding cycle. One such unit is the column and boom shown alongside the positioner in Fig.2.33. The boom is adjustable for height on the column and is provided with a variable speed drive laterally. Thus a cylindrical workpiece secured in the chuck attached to the faceplate of the positioner can be surfaced on its diameter with a continuous spiral weld bead as the manipulator traverses the welding head along the workpiece.

Sometimes, it may be desirable to weave the weld bead and to achieve this an oscillating mechanism is attached to the end of the boom. An alternative form of manipulator is shown in Fig.2.33; this is a side beam carriage and provides linear movement only. In this example, the side bead is teamed with a lathe-type manipulator designed to rotate long shaft type workpieces between a chuck and back centre. To cater for variations in the shaft diameter, an arc length control system is fitted between the torch and the carriage.

2.34 Side beam carriage manipulator and lathe type positioner for PTAW surfacing of shafts (courtesy Deloro Stellite Ltd).

2.35 Fully mechanised and programmed surfacing system for production line work (courtesy Deloro Stellite Ltd).

The equipment illustrated in Fig.2.33 and 2.34 could be used for TIG, PTA or MIG/MAG and of the two, the column and boom set-up would provide the more flexible arrangement for a shop handling a variety of work.

Figure 2.35 shows a purpose built unit for incorporation in a production line. It is designed to weld surface small circular components such as valves and seats or seal rings. Built to use the PTA process, it incorporates an enclosed welding cabinet which contains a small tilting and rotating table to carry the workpieces. The PTA torch is carried on the end of an arm extending out through the cabinet wall to the oscillating mechanism fitted to the left side of the machine which provides weld bead weaving when needed. The weld consumable is an alloy powder held in the container on top of the cabinet and fed through a metering device to the torch. On top of the welding power supply is a system which controls the whole welding cycle to preset parameters. This can include mechanised loading and unloading of the workpieces through the rear of the cabinet.

The advantage of such a system is that once weld surfacing parameters have been set, the unit operates with minimum attention provided that supplies of workpieces and surfacing consumables are maintained. Provision can be made to store a series of programs to suit a range of workpieces.

A large number of weld surfacing operations are carried out on components with flat or cylindrical surfaces, which require only simple positioning and manipulation and are welded in the downhand position. More complex surfaces, such as the screw segment shown in Fig.2.36, which is part of an oil expeller used in the food industry, are ideal for using a robot, particularly when teamed with the positional welding capability of the MIG/MAG process, as this enables the complex shape to be coated all over without intermediate repositioning of the workpiece and welding head.

ACCESSIBILITY

The problem of access arises in weld surfacing when the region to be deposited is located inside a component in such a way that it is difficult to get the welding torch and consumable feed inside and present it at the required angle to the work surface. Even if this is possible, it may be difficult to see what is happening and make suitable adjustments to the welding parameters while deposition is taking place.

In such circumstances, manual deposition is usually not possible because of poor visibility and consequent loss of control. The difficulty can be avoided

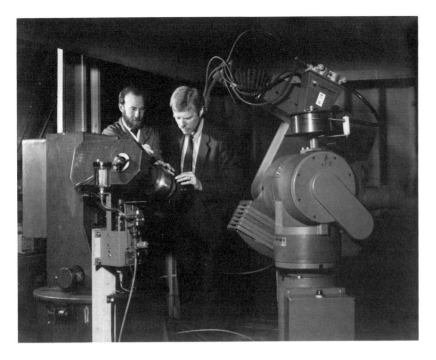

2.36 Robotics being used for weld surfacing expellor screw segments to resist wear by abrasion and corrosion in the food industry (courtesy Simon-Rosedowns Ltd and Taylor Hitec Ltd).

by fabricating the piece after deposition of the temporarily accessible surface; welding or brazing castings of the surfacing alloy in position or melting precast shaped inserts in position by means of a modified TIG torch of slender proportions.

Two recent developments have, however, made this type of operation easier for designers. The first is the introduction of small PTA torches, which work with a powder consumable and are capable of working in a bore of 35mm, Fig.2.37, with a nozzle set at 90 or 45° to the torch axis. Current models have a 'reach' of about 450mm. A longer reach is possible using specially designed MIG/MAG guns, but in this case, minimum diameter is about 75mm. Both types of torch/gun are designed for mechanised use only and it is normal to secure area coverage of the deposit by arranging for a slow spiral deposit with suitable bead overlap. This requires a suitable speed relationship between the positioner and manipulator. To obtain satisfactory results, the component must be true and concentric to the axis of rotation. By turning end to end, it is possible to cover a bore double the length of the torch reach, assuming access from both ends.

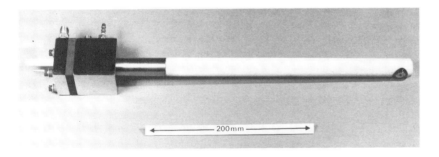

2.37 PTA torch designed for use inside bores of 35mm diameter (courtesy Deloro Stellite Ltd).

WELD DEFECTS

The specification under which work is carried out should not only lay down the surfacing procedure to be used and precautions to be observed but should define tests and acceptance standards to be met. If the finished parts, on inspection, fail to meet these standards, it is essential to identify faults which may be rectified, and the rectification procedures which are permissible. The inspection schedule may include in-process inspections at various stages of the work — some are obvious, such as the cleanness of the blank, adherence to dimensional tolerances, thickness of deposit, deslagging between successive deposit layers, and so on. This section is confined to comments on inspection after surfacing has been completed. The designer can use these in development of a specification suited to each job.

Inspection is facilitated if the work, after weld surfacing, is grit blasted to remove scale and oxides. Grit blasting after rough machining is a valuable test for soundness in the deposit as it reveals porous areas. Proof machining of components which have to be finally ground or machined to precise dimensions is essential, as inspection at this stage reveals undersize areas which can be built up further while there is sufficient material in the rest of the part available to cover scaling or distortion, as noted earlier. The rectification allowable depends on the process and alloys used and can range from spotting or more extensive additional surfacing, observing the necessary heating requirements, to complete removal of the deposit and re-initiation of the surfacing cycle. A repair procedure should be laid down rather than be left to the operator.

The deposit must be viewed carefully for cracks, porosity, pinholes and inclusions which may have to be rectified. Visual inspection may be aided by dye penetrant inspection.

When used on an as-welded surface care must be taken in interpretation of dye penetrant tests. Such tests should be repeated after proof machining. At this stage additional tests may be specified, such as hardness, radiographic or ultrasonic examinations. As surface cracking may be caused by incorrect grinding of harder deposits after final machining dye penetrant testing should be repeated, as well as dimensional checks.

The results of a testing procedure can vary according to the skill of the operator and the techniques which are used. It is therefore desirable that techniques be laid down in detail, and audited from time to time. An acceptance standard should be stated for each job.

Cracks

Contraction cracking
Contraction cracking is normally visible to the naked eye and may not necessarily affect the performance of the component adversely. In the harder surfacing materials providing maximum wear resistance, such cracking (relief cracking) may be encouraged to release locked-in tensile stresses. Such cracking rarely involves the risk of the coating breaking away from the base metal, provided there is no hardening of the HAZ, and a satisfactory bonding to the substrate has been achieved. However, cracks cannot normally be tolerated in:

1 Sealing surfaces of valves, mechanical seat rings, printing rolls, *etc*;

2 Surfaces subject to erosion, such as flow control valves;

3 Surfaces designed to provide both wear and corrosion resistance;

4 Surfaces subject to severe fatigue stresses in service;

5 Surfaces which must not pick up any process material which could contaminate subsequent batches, such as in plastic extrusion.

Grinding cracks
Grinding cracks are usually extremely shallow but as they may extend under mechanical or thermal stress they must be treated as contraction cracking.

Sub-surface cracks
If the deposit lifts from the base metal areas may flake off or spall even under slight mechanical pressure or relatively small surface temperature change and therefore such parts must be rejected, not repaired.

Porosity
The degree of porosity, and the size of the pores, can vary enormously. There can be a few specks invisible to the unaided eye to large pores and visibly spongy areas if poor preparation or technique has been used. Pictorial standards are desirable from which a grading can be specified for each job. A numerical approach is an alternative, e.g. a maximum of X pores per square Y, the pore size not to exceed Z microns. The designer should not call for a higher quality than is required for the efficient performance of the surface, and this may differ from one area to another. Thus with a seal or valve seat the hardfacing frequently extends beyond the actual sealing zone and some porosity outside this zone may not be detrimental. The component drawing

should show the different zones and indicate the acceptable standard for each.

Inclusions

Inclusions, like porosity, can range from visibly obvious defects to very fine dispersions. Removal on machining or during the working life of the part produces the same problems as pores. Their identification and classification can be treated similarly.

TREATMENT OF DEFECTS

Cracks

For applications where cracking of any sort is unacceptable, it is sometimes possible to remove the entire deposit and to start again, provided that:

— The cause of cracking is known and recurrence can be prevented;
— There is adequate machining allowance on uncoated areas of the component;
— The part, after removal of the faulty deposit, is heat treated to remove welding stress and to correct any unacceptable metallurgical structures, such as a hard HAZ;
— The increased thickness of deposit resulting from machining to remove the faulty deposit and to get down to clean, uncracked substrate material, is acceptable from the design and use point of view and unlikely to cause recurrence of the cracking problem;
— The newly prepared blank is checked by dye penetrant testing and found to be free from all defects.

Once formed, grinding cracks in some materials, although initially shallow, propagate more deeply if further stock removal takes place; so total removal may sometimes be necessary before rewelding.

Any defects penetrating locally into the blank should be excavated by local grinding to give an open recess, not a drilled hole, so that the welding flame or arc can access the bottom of the hole. The defect can then be rewelded, stress relieved, remachined and tested before resurfacing. Never use the surfacing alloy to fill excavations as the local change in deposit thickness can cause further cracks.

Do not attempt to repair cracks in the deposit or the substrate without excavation to remove the defect entirely; the heat of welding is likely to propagate the defect even more deeply.

Porosity and inclusions

The distribution of porosity, if unacceptable, determines the method of rectification. A localised defect possibly in the final weld crater may be repaired by local excavation and rewelding after dye penetrant testing to verify complete removal of the defect. As with cracks, attempts to remove defects by welding directly invariably fail.

More widely distributed defects suggest a fundamental problem with the materials used or the procedures and this will require investigation. If the cause is identified and can be avoided in future, it is probably best to remove the whole deposit and reweld.

Inclusions should be treated in the same way as porosity.

The case for and against repair of defects, assuming that it is considered to be a practical proposition, depends on factors such as the cost compared with using a new blank, whether sufficient machining allowance is still available on uncoated areas and on metallurgical considerations of effects of further exposure to welding temperatures on the substrate.

Explosive weld cladding and friction surfacing

Explosive weld cladding and friction surfacing allow application of a variety of materials to the surface of cheaper substrates without the coating material being raised to the molten state. Bond strengths approach those of conventionally welded surfaces without problems of dilution.

EXPLOSIVE WELD CLADDING

It was well known during the First World War that a bullet or shrapnel could adhere to the metal surfaces they impacted. This scientific phenomenon was not appreciated as a welding process at that time and it was only in 1944 that Carl recognised it as a solid phase welding process. This chance observation has been developed during the past 45 years to what is today a well established industrial process.

Although explosive welding (like any other process) suffers some limitations, it is one of the most effective techniques for welding dissimilar metals. It is generally used for production of composite metal sheets and for lining tubular geometries with thin layers of alloy materials for corrosion protection.

Theory of the process

Explosive welding is achieved when a metal plate is accelerated by an explosive charge to a high velocity oblique collision with another metal plate. Figure 2.38 demonstrates the basic configuration of explosive welding where a flyer plate is accelerated under the influence of the detonation gas to impinge obliquely on to a base plate. In doing so a jet of metal (from both plates) in the form of a spray is ejected ahead of the collision point cleaning the weld surfaces of any oxide films or surface impurities. Pressure at the collision point is estimated at 250Kbars which is well in excess of the yield strength of any metal. Thus, under these conditions the interfaces of the two metals are subjected to an interatomic contact where the cohesive energy or more simply the balance of the interatomic forces between the metal atoms results in a solid phase weld.

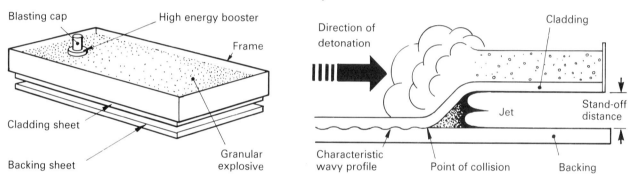

2.38 *Explosive welding where a flyer plate is accelerated by detonation gas on to a base plate.*

Stainless steel, nickel, titanium and other metals are frequently chosen for their corrosion resistance, but when used in the solid state, most of their thickness is required for physical strength. As only the exposed surface, plus a few millimetres, (depending on the corrosion/erosion rate) is required to give protection, considerable savings can be made by using the expensive metal where it is needed and a less expensive one to provide strength.

Explosive welding can clad a wide range of metals to almost any backing. The process offers several important advantages:

— Up to 100% of area bonded;
— A metallurgical bond which is stronger than the weaker of the two metals;
— There is no HAZ for many combinations;
— Butt welding of the finished product is possible;
— The backing metal can be double clad with a different material on either side.

Practical applications

The main applications of explosive welding are in flat plate cladding and welding or cladding of cylindrical geometries. Explosively clad plates are widely used in the following sections of industry:

1. Stainless steel/aluminium composites are used in production of freezer trays for the fast freeze food industry.

2. Thick steel plates are clad with a thin layer of corrosion protection materials such as titanium, monel and nickel alloys. These plates are used for fabrication of storage tanks for process plants.

Explosive cladding of cylindrical surfaces was an early development in explosive welding. Initially it was applied in particular to nozzles in steam power plants for corrosion protection. Recent developments on new explosive compositions and production clamp designs have produced commercially viable components, such as long weld neck flanges clad with nickel alloy materials, Fig.2.39, spool pieces and flowlines clad with nickel alloy 625, Fig.2.40 for subsea production systems. A major contribution to the pipeline welding industry is explosive welding of stainless steel duplex pipe with both practical and commercial advantages in J laying of pipelines for the offshore oil industry.

Further, explosive welding can be controlled so it can be applied to production of microwelds between metal foils as thin as 11 µm. An example of this is coating of titanium sheets with platinum.

Explosive welding has and can be used in other areas of industry but the practicality and the commercial viability of each system must be assessed on its own merits.

2.39 *Long weld neck flange clad with nickel alloy (courtesy Hotforge Ltd).*

2.40 *Flowline clad internally with nickel alloy 625 (courtesy Hotforge Ltd).*

FRICTION SURFACING

In 1941 Klopstock and Neelands filed a UK patent application for 'an improved method of joining or welding metals'. The information contained in the patent introduced the concept of depositing metals by friction surfacing. More recently, however, there has been renewed interest in this method of deposition which shows promise for a wide range of applications. Essentially friction surfacing is a solid phase deposition technique whereby a rotating consumable bar usually in the range 10-40mm diameter is pressed on to a laterally moving substrate, Fig.2.41.

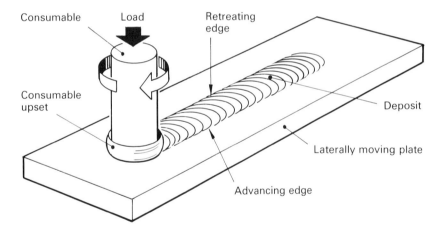

2.41 *Basic principle of friction surfacing.*

Friction surfacing has the advantage of allowing materials to be used which would otherwise be metallurgically incompatible with the substrate, yet still providing a layer of deposited material with a high degree of adherence. Special combinations of material properties that cannot be realised in monolithic materials can be provided which conserve more expensive and strategic materials.

It has been established that corrosion problems represent approximately 4% and wear itself has been estimated at over 1% of the gross national product. Often it is the combined tribological effect of wear and corrosion that is so damaging. These enormous losses can be reduced. Friction surfacing offers an alternative to many other techniques, and can be used to combat these problems.

Metal combinations and properties

The deposit, a product of a hot forging action, is inherently homogenous and of good mechanical strength, see Table 2.3. The solid phase nature of the process ensures that negligible dilution is achieved yet good basic adhesion properties are maintained. The interface region usually remains intact even after resisting loads even to the ultimate tensile strength of the weaker material.

Table 2.3 Details of surfacing conditions, deposit thickness and mechanical test results

Materials		Welding machine settings						Mechanical tests TTT*
Deposit	Substrate	Consumable size, mm	Rotation speed, rpm	Friction force, kN	Substrate traverse, mm/sec	Touchdown period, sec	Average deposit thickness, mm	Tensile strength, N/mm²
Mild steel	Mild steel	25	975	28	4	3	1.9	510
Mild steel	Low alloy steel	25	975	28	4	3	2.10	
Austenitic stainless steel	Mild steel	25	550	50	5.3	3	1.30	570
Austenitic stainless steel	Austenitic stainless steel	25	550	50	4	4	1.50	537
Alloy 625	Mild steel	20	410	45	4	1.5	1.30	470
Hastelloy CW-12M-1	Austenitic stainless steel	20	330	56	4	2.0	1.30	433
Stellite grade 6	Austenitic stainless steel	20	330	39	5	5.0	0.70	620
Aluminium alloy (2011)	Aluminium alloy (20.4A)	25	778	17	4.2	3.0	3.10	237

*TTT = through-thickness tensile

With respect to composite clad layers, corrosion resistant and hardfacing materials usually have superior mechanical properties compared with the substrate material. It is not unexpected, therefore, that for optimised deposits, subjected to through thickness tensile testing, that failure occurs at loads equalling the ultimate tensile strength of the substrate. Lateral restraint, provided by a stronger deposit material, to the interface region usually means that failure occurs well away from the joint. With like to like deposit/substrate materials failure sometimes occurs in cohesion within the bulk deposit. With relatively weaker deposit materials, failure usually occurs in cohesion of the bulk deposit.

Comparative corrosion resistance properties established from potentiodynamic polarisation curves for stainless steel deposits in chloride and acidic solutions, between machined deposits and consumable specimens indicated that corrosion resistance properties are maintained. Pin-on-disc comparative wear tests between cobalt base alloys (Stellite grade 6) have shown that the deposit wear properties at least equal those of the consumable.

GENERAL CONSIDERATIONS
During initial dynamic contact between the rotating consumable and the substrate the resultant relative motion under an axial load produces a scouring action which removes the oxide layer from both contact faces. After the initial contact and after traverse has been initiated the scouring action necessary to disperse the intrusion of the substrate oxide barrier is continued, not by the contact face of the consumable directly, but by the plasticised layer produced from the consumable. While it is recommended that the plasticised material has lower mechanical strength it is expected that provided suitable axial force is maintained oxide dispersal continues, resulting in sound bonds.

The unequal temperature distribution between the comparatively small consumable bar and the bulk substrate leads to preferential consumption of the bar. This asymmetrical temperature distribution is further enhanced during friction surfacing by continually introducing new substrate at ambient temperature.

Although regions of the rotating consumable are repeatedly exposed to the atmosphere, sound deposits of good integrity can be produced. More reactive material, however, may benefit from a suitable shielding gas.

Physical mechanisms
Lack of symmetry characterises the relative motion and material transfer of the process. One observable characteristic, more noticeable at lower rotational speeds, is the slight difference in appearance between the two side edges. Figure 2.42 shows that relative movement between the bar and

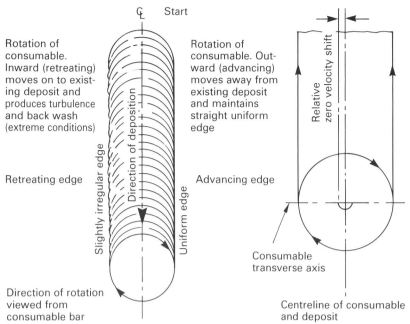
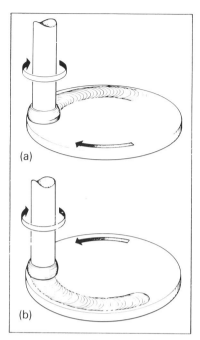

2.42 Deposit characteristics.

2.43 Friction surfacing annular deposits: a) Consumable and substrate, offset, with same angular direction; b) Consumable and substrate, offset, with opposing angular direction.

the substrate is enhanced on the forward or advancing and conversely is diminished on the retreating side. The introduction of lateral motion of the substrate to the angular rotation of the consumable produces asymmetric effects during deposition. Annular deposits produced with the consumable and substrate revolving in the same angular direction, Fig.2.43, nominally minimise this asymmetric effect especially with increasingly larger diameter consumables.

The uniform surface ripple, Fig.2.44a, results from the interaction of rotation and substrate traverse. Any point towards, and including, the periphery of the bar, moves in a cycloidal path, known as a compact superior trochoid. The almost semi-circular ripples on the deposit surface point towards the start of the run and are produced by the final sweep of the trailing edge of the rotating consumable. Investigation has shown that at the substrate surface the scouring marks point towards the stop end of the deposit. Thus a complete reversal of relative motion must take place, Fig.2.45, within the central zone of the deposit thickness. However, the extreme edges of the deposit only experience one direction with respect to the rotating consumable, *i.e.* one side retreating and the other side advancing with respect to the traverse.

Unequal heat generation

The effect of the relative heat generation can be seen in any transverse macrosection. Figure 2.44b shows a shallow crescent-shaped HAZ of maximum depth in the centre, insufficient heating and lack of bond at the outer edges of the deposit are also shown. Tubular consumables as well as shaping of the substrate modify this effect.

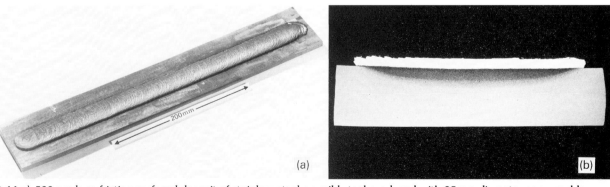

2.44 a) 500mm long friction surfaced deposit of stainless steel on mild steel produced with 25mm diameter consumable; b) Transverse macrosection of deposit, consumable speed 650rpm, force 62kN, traverse rate 7.5 mm/sec.

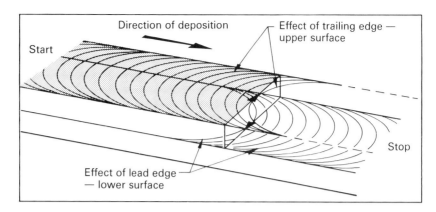

2.45 *Motion on the upper and lower surfaces of the deposit.*

Lack of bonding at the outer edges of the deposit is inherent to operating on flat substrates. The outer edges of the deposit only receive action at tangential points corresponding to the outer diameter of the bar. On the other hand in the centre zone the full bar acts upon the plasticised layer and deposited material for a longer period. In addition, there is not the same degree of pressure at the outer edges of the consumable bar, and this contributes to the unequal distribution of frictional effort.

APPLICATIONS

Friction surfacing has already been used for reconditioning of worn shafts and good quality deposits have been reported. Successful hardfacing of cutting edges and agricultural tools has also been carried out by Soviet workers.

Interest in the process continues and as a result applications, mainly in localised cladding are being investigated. Various substrate geometries are shown in Fig.2.46. Typical examples include the anti-corrosion overlay of slide valve plates; surfacing of the annular contact face of composite pipe flanges, disc brakes, *etc*, Fig.2.47*a*; inlay cladding of strategic materials positioned to suit bearings or 'seal' contact areas on shafts, Fig.2.47*b*, and

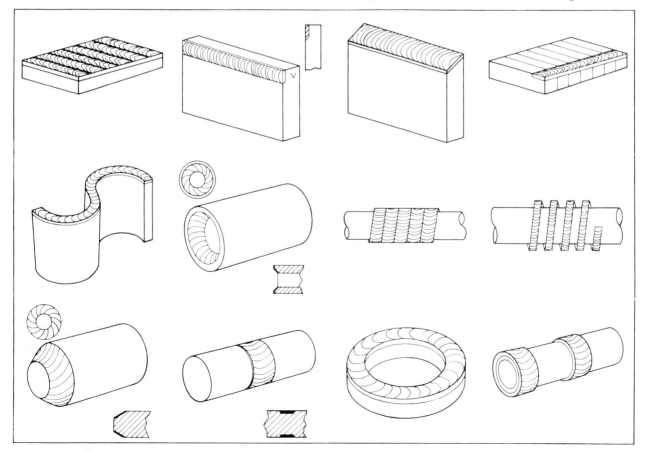

2.46 *Suggested geometric arrangements for friction surfacing.*

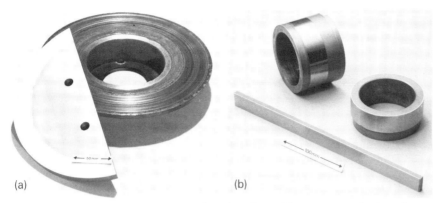

2.47 *Friction surfaced components produced at The Welding Institute:* a) *Bimetal brake disc inlayed with wear resisting material;* b) *Tubular mild steel shaft inlayed with stainless steel and tool steel clad layers on to flat and tubular components.*

2.48 *A wire rubbing pad inlayed with wear resisting materials which has greatly improved the surface life of the component (courtesy Friction Technology Ltd).*

cladding of the exposed regions of shafts that see service in pernicious environments. Further suggestions include manufacture of specialised wear tiles for sinter plant, wire rubbing pads, Fig.2.48, quarrying equipment, bi-metal linear bearings and also in-situ reclamation of worn railway points. Turbine and guillotine blades are also considered examples of suitable applications for friction surfacing.

The process will increasingly become concomitant with that of other solid phase large overlay techniques. Repair and reclamation of materials already in clad form can also be carried out. The Welding Institute considers that artillery shell banding may be another application area, where suitably soft material can be friction surfaced to shell bodies.

Advantages
1 The solid phase friction surfacing process minimises distortion.

2 Because of its inherent fast freeze characteristic friction surfacing allows deposits to be made in all positions.

3 Thin layers of negligible dilution but considerable bond strength can be obtained for a range of corrosion and wear resistant materials.

Disadvantages
1 Unless the consumable is wider than the substrate component, or deposition is to be made into a recessed groove, lack of bonding at the outer edges of the deposit is noticeable. This distracting feature is an inherent effect which reduces the effective bond width to just less than the diameter of the consumable.

2 The substrate needs to be able to react to high compressive loads; internal support may therefore be necessary for thin walled hollow components.

For engineering coatings of all types - ask METCO

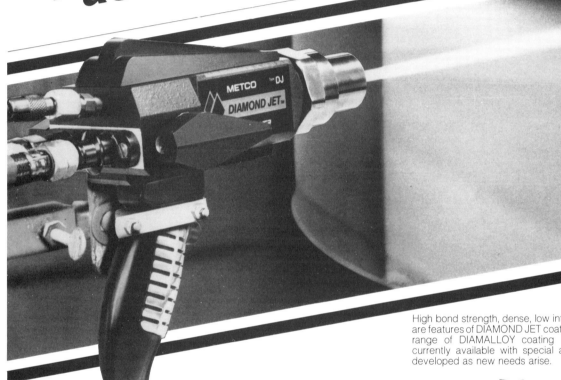

High bond strength, dense, low internal stress are features of DIAMOND JET coatings. A wide range of DIAMALLOY coating materials is currently available with special alloys being developed as new needs arise.

As world leaders in all thermal spray processes, METCO will give unbiased advice for the right process to suit your particular needs. From aluminium coatings against corrosion on steelwork to zironia as a thermal barrier. From simple, hand operated spray guns for only occasional use to pre-programmable computer control of METCO high-tech spraying equipment and robotic work-handling installations.

And now from METCO comes the new DIAMOND JET system. This High-Velocity Oxygen-Fuel (HVOF) process introduces a new generation of thermally sprayed coatings.

Details of METCO thermal spray processes and over 150 coating materials are described in this literature. Please ask for copies.

METCO LTD Chobham Woking Surrey GU24 8RD
tel: Chobham (0276) 857121 fax: (0276) 856307 telex: 858275 METCO G

Thermal spraying processes and materials

CHAPTER 3

The principal characteristics of thermal spraying processes which distinguish them from weld deposited coatings are as follows:

- Adhesion to the substrate. The strength of the bond between coating and substrate covers a wide range, depending on the materials and process used. It varies from relatively low strengths to figures approaching those of welded bonds if the process involves high temperature diffusion between coating particles and substrate.
- Thermal spraying can apply coating materials to substrates which are unsuited to welding because of their composition or tendency to distort. This feature offers the designer scope to use materials with desirable properties which would not be possible by other means.
- Sprayed deposits can be applied in thinner layers than welded coatings, but thick deposits are possible under certain circumstances.
- Almost all material compositions may be deposited (provided there is a stable liquid phase), metals, ceramics, carbides, plastics or any combination. This chapter deals with coatings other than plastics which are covered in Chapter 6.
- Most processes are 'cold' compared with welding and there is no distortion, dilution or metallurgical degradation of the substrate.
- Thermal spraying processes are all-positional and can be operated in air, so they possess great flexibility.

In all thermal spraying processes, the consumable coating material fed to the spray gun is raised in temperature and projected in particulate form to strike the workpiece. On arrival, the hot particles form splats which interlock and gradually build up a coating of the desired thickness, Fig.3.1. The density of this coating depends on the material, its temperature when its strikes the workpiece and its impact energy. Adhesion of the coating to the substrate depends on the same factors plus the surface condition of the substrate which must be clean and suitably roughened.

Processes available for thermal spraying have been developed specifically for the purpose and fall into two categories. The lower energy processes, often referred to as metallising, are arc and flame spraying. These are extensively used for spraying metals for corrosion resistance such as zinc and aluminium, for service at or near ambient temperature, on large structures and in circumstances where thermal and mechanical shock or abrasive wear are small. Some porosity will always be present in these coatings; it may be beneficial, as residual stress in coatings tends to be lower and thicker coatings can be applied without risk of lifting from the substrate. Porosity can also act as a reservoir for lubricant for a sprayed shaft running in a bearing. Porosity can be sealed in circumstances where corrosion could otherwise penetrate through it to the substrate.

The higher energy processes, plasma, D gun and high velocity combustion spraying were developed to provide coatings of much lower porosity and improved adhesion to the substrate and also to handle materials of higher melting points, thus widening the range of application to include resistance to higher temperatures and to thermal and mechanical shock.

With lower energy processes, adhesion to the substrate is considered to be largely mechanical and is dependent on the workpiece being perfectly clean and suitably rough. Roughening is carried out after preparation machining by shot blasting with angular iron grit or by rough machining (e.g. threading).

With higher energy processes, bond strengths are higher because of the disruption of any oxide films present on the particles or the workpiece surface. When this occurs some diffusion bonding takes place. However, cleaning and roughening of the workpiece surface is just as important as for the lower energy processes if best results are to be obtained consistently. Aluminous grits are normally used in place of angular iron grit.

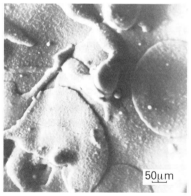

3.1 The surface of a plasma sprayed oxide. Note how the molten droplets have splat and spread (courtesy Materials Engineering Centre, AERE).

77

Other differences distinguish the two groups of processes mentioned above. The low energy 'metallising' group has lower capital cost and is relatively portable, while the higher energy group has a higher capital cost and tends to be used in fixed installations. However, there are some industrial applications where a process from either group could be used successfully.

Included in this chapter is the spray fuse process. This technique involves spraying to apply a coating of special self-fluxing alloys based on cobalt or nickel. The coating is then raised in temperature to between the solidus and liquidus of the alloy, when the self-fluxing action allows diffusion bonding between the sprayed particles and the substrate. Fusing can be effected in a number of ways.

Consumables used for thermal spraying are available in two forms. Solid drawn wire is used for metallising processes and some materials are produced as a dispersion of particles in an organic binder, which is also in wire form and is known as a 'cord'. For higher energy processes a powder consumable is used. This has advantages for non-metallic materials and can also be transported more easily to the point of spraying, especially for torches designed to be used in small bores.

Thermal spray materials

MATERIALS WITH GOOD BONDING CHARACTERISTICS

Experience has shown that certain products provide a better bond to the substrate than coating materials themselves. This fact has led to their use as an interlayer and also in the form of a blend with some surfacing alloys. These are listed in Table 3.1, and are commonly called bond coats.

Table 3.1 Bond coats

Materials	Characteristics and applications	Suitable for (processes)					
		Oxyfuel gas wire	Oxyfuel gas powder	Fused coatings	Electric arc spray	Plasma spray	Detonation coating
Nickel-aluminium	95-5 and 80-20 composite and pre-alloyed materials	√	√			√	
Nickel-chromium-aluminium	As for nickel-aluminium but giving higher temperature resistance		√			√	
Molybdenum	Useful as a wear-resistant coating	√	√		√	√	
Ni-Cr-B-Si/Ni-Al Ni-Cr-B-Si/Mo	Typical of a number of blended powder materials			√			

For use as a bond coat, spraying parameters for molybdenum are chosen to minimise oxidation in the deposit. The hardness of the deposit rises with an increase in oxide content, producing useful wear resistant properties. For example, deposits are applied to the synchronising rings in gear boxes, a typical hardness for these being 650HV. Hardnesses in the range 400-800HV are possible with suitable spraying parameters. These harder deposits are resistant to galling and fretting and are used, for example, on engine piston rings. Maximum temperature of use is 300°C.

FERROUS ALLOYS

Mild, low alloy and carbon steels are most frequently used to build up worn areas of components where corrosion resistance is not required, Table 3.2. Stainless steels can also be sprayed and are frequently used for reclamation where corrosion resistance is needed.

NICKEL BASE ALLOYS

All nickel base alloys are corrosion resistant and those containing chromium have good resistance to oxidation at elevated temperatures, Table 3.3. Nickel-chromium-boron-silicon alloys can be fused after spraying to give coatings which are free from porosity and are metallurgically bonded to the substrate.

A range of alloys developed for resistance to high temperature oxidation and sulphidation, particularly in gas turbine engines, has found application in industry. These are referred to as the M-Cr-Al-Y series, where M can be Fe, Co, Ni or Ni-Co.

Table 3.2 Ferrous alloys

Steel type	Characteristics and applications	Suitable for (processes)							
		Oxyfuel gas wire	Oxyfuel gas powder	Fused coatings	Electric arc spray	Plasma spray	Detonation coating	Jet Kote	Diamond Jet
Mild	Build-up coat before application of harder materials; readily machinable	√			√				
Carbon and low alloy	Reclamation of worn parts not subject to corrosion in service; finished by grinding	√			√				
Martensitic	Moderate corrosion and high wear resistance, best finished by grinding	√	√		√	√			
Austenitic stainless	High corrosion resistant coatings, impact resistant, good machinability	√	√		√	√		√	√

Table 3.3 Nickel alloys

Alloy type	Characteristics and applications	Suitable for (processes)							
		Oxyfuel gas wire	Oxyfuel gas powder	Fused coatings	Electric arc spray	Plasma spray	Detonation coating	Jet Kote	Diamond Jet
Nickel-copper	Gives high density machinable coatings with good corrosion resistance, especially under marine conditions	√	√		√	√			
Nickel-chromium-molybdenum-tungsten	Resistant to corrosion and wear; work hardens under impact					√	√	√	
Nickel-chromium-boron-silicon	A range of alloys with hardnesses from 200-700HV. High wear and corrosion resistance and good retention of hardness at temperature		√	√		√	√	√	√
Nickel-chromium-boron-silicon-copper-molybdenum	High corrosion, wear and heat resistance with wide fusing range enabling it to be used on complex shaped components		√	√		√			
80-20 nickel-chromium	Resistant to corrosion and high temperature oxidation. Used as a precoat for application of ceramics	√	√		√	√	√	√	
Nickel-chromium-iron	Corrosion and high temperature resistant coatings	√	√		√	√			√

COBALT BASE ALLOYS

Alloys based on cobalt have high resistance to softening at elevated temperatures together with corrosion and wear resistance, Table 3.4. Alloys modified by inclusion of boron and silicon can be fused after spraying to give pore-free metallurgically bonded overlays (see spray fusing later in this chapter).

TUNGSTEN CARBIDE AND CHROMIUM CARBIDE

As a sintered composite with cobalt, tungsten carbide is used for plasma and detonation coatings, Table 3.5. It blends with nickel or cobalt base self-fluxing alloys and can be used to provide fused coatings. Chromium carbide is used, generally in conjunction with Ni-Cr, to confer resistance to high temperature wear.

CERAMICS

Ceramics have high temperature resistance and are resistant to wear and erosion, Table 3.6. Coatings sprayed by oxyfuel gas processes from rods or

Table 3.4 Cobalt alloys

Alloy type	Characteristics and applications	Suitable for (processes)						
		Oxyfuel gas wire	Oxyfuel gas powder	Fused coatings	Electric arc spray	Plasma spray	Detonation coating	Jet Kote
Cobalt-chromium-tungsten	A range of alloys with hardnesses from 300-600HV. High resistance to wear, corrosion and heat					√	√	√
Cobalt-chromium-nickel-tungsten	Used for coatings where heat resistance combined with resistance to fretting wear is required					√	√	√
Cobalt-chromium-molybdenum	Hardness 300HV, this alloy has higher corrosion resistance and ductility than the above					√	√	
Co-Cr-Mo-Si intermetallic						√		√
Cobalt-chromium-tungsten-boron-silicon	A range of alloys with hardnesses from 400-700HV. Used for production of fused coatings			√		√		

Table 3.5 Carbides

Materials	Characteristics and applications	Suitable for (processes)							
		Oxyfuel gas wire	Oxyfuel gas powder	Fused coatings	Electric arc spray	Plasma spray	Detonation coating	Jet Kote	Diamond Jet	CDS
Tungsten carbide/cobalt	Composite powders with 10-20% cobalt. Used for coatings subject to wear plus impact at temperatures up to 450°C					√	√	√	√	√
Tungsten carbide/Ni-Cr-B-Si alloy blended powders	Fused coatings having very high wear resistance	√	√			√			√	
Tungsten carbide/Co-Cr-W-B-Si alloy blended powders	Fused coatings having very high wear resistance	√	√			√				
Chromium carbide 80%Ni-20%Cr blended powders	Composite powders for coatings subject to wear at temperatures from 500-1000°C					√	√	√	√	√

Table 3.6 Ceramics

Materials	Characteristics and applications	Suitable for (processes)						
		Oxyfuel gas wire	Oxyfuel gas powder	Fused coatings	Electric arc spray	Plasma spray	Detonation coating	CDS
Alumina	Wear resistant coating with high electrical resistance	C	√			√	√	
Zirconia	High thermal insulating coatings having good resistance to thermal shock	C	√			√		
Chromic oxide	Very high wear resistance and resistance to abrasion and cutting by threads, etc	C	√			√		
Alumina-titania	A range of compositions from 2-40% titania; coatings give higher wear resistance than 'pure' alumina	C	√			√	√	√
Titanium dioxide	Wear and high temperature resistant coatings; good corrosion resistance	C	√			√		
Magnesium zirconate	Thermal barrier coating with high resistance to erosion by fine particles and molten metals		√			√		

C = available as plastic covered cord

cords, or from powders applied by plasma or detonation, have high density, but flame spray coatings made using powders may be porous and may sometimes need to be sealed. Porosity is sometimes deliberately sought to give higher thermal insulating properties. Many proprietary blends of ceramic powders are used, but those given in Table 3.6 are considered the most important.

Thermal barrier coatings are used in equipment operating at high temperatures to reduce the temperature of the underlying metal so that it lies within its limits of operation. This allows higher gas temperatures and therefore greater operating efficiency in gas turbine and automotive engines.

Examples of thermal barrier coating systems are shown in Fig.3.2 and 3.3.

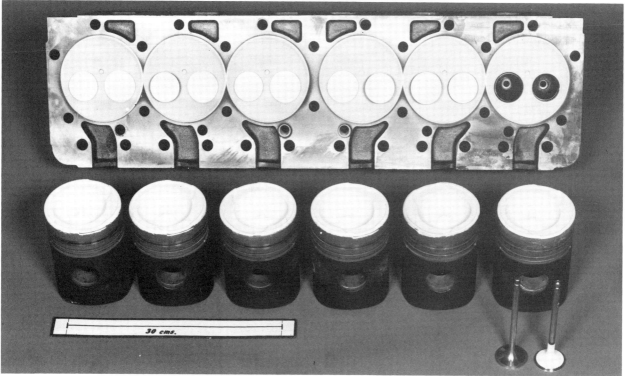

3.2 *Future diesel engines may be insulated for improved efficiency. Components of a diesel engine coated with a 2mm thickness thermal barrier applied by plasma spraying (courtesy of Materials Engineering Centre, AERE).*

MISCELLANEOUS METALS AND ALLOYS
Other metals and alloys such as aluminium and zinc find specific application where properties such as resistance to atmospheric corrosion are of prime importance, Table 3.7.

Table 3.7 Miscellaneous metals and alloys

Materials	Characteristics and applications	Suitable for (processes)					
		Oxyfuel gas wire	Oxyfuel gas powder	Fused coatings	Electric arc spray	Plasma spray	Detonation coating
Aluminium	Used principally as a corrosion resistant coating on steels. When heat treated provides a high temperature oxidation resistant alloy coating on steels	√	√		√	√	
Zinc	Used exclusively for corrosion resistant coatings on steels	√	√		√		
Bronze	For bearing surfaces	√	√				
Aluminium bronze	Corrosion resistant coatings with good bearing properties. Also used as a bond coat with arc spray process	√	√		√		
Copper	Used mostly for its electrically conducting properties	√			√		

CERMETS AND GRADED COATINGS

Because of wide differences in the expansion characteristics of ceramics and steels it is sometimes necessary to use a cermet or graded coating system to reduce the tendency to cracking and spalling under thermal cycling. In these situations it is often possible to use a first coating with 70-80% metal component, a second layer of 20-30% metal, and a final layer wholly of the required ceramic.

Cermets are useful in other circumstances; for instance, to combine wear resistance of a ceramic material with the higher thermal conductivity of a metal.

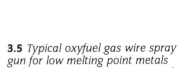

3.3 Microstructure of a typical plasma sprayed three layer thermal barrier system — metallic bond (NiCrAl), cermet (NiCrAl + ZrO_2), zirconia (A = substrate, B = NiCrAl; C = cermet; D = zirconia) (courtesy of Materials Engineering Centre, AERE).

Characteristics of thermal spray processes

OXYFUEL GAS WIRE SPRAYING

In oxyfuel gas wire spraying the coating material, in the form of rod, wire, or cord, is fed by a variable speed motor into the centre of the multijet flame which melts the tip of the wire; an annular gas jet then propels the molten material on to the substrate, Fig.3.4 and 3.5.

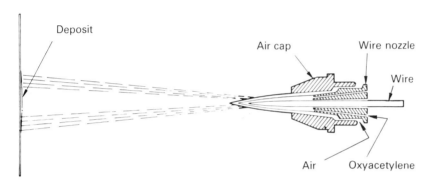

3.4 Wire combustion spray method.

3.5 Typical oxyfuel gas wire spray gun for low melting point metals including zinc, aluminium, tin and Babbitt (courtesy Metco Ltd).

Advantages
1 Relatively low cost equipment;
2 High spray rates;
3 Wide range of surfacing materials can be sprayed;
4 Adaptable for differing wire diameters;
5 Spraying can be mechanised.

Disadvantages
1 Limited to consumables in wire form;
2 Lower density deposits and lower adhesion strength than arc spray.

Principal applications
Wire spraying is used extensively for deposition of zinc and aluminium coatings for corrosion protection. Worn or mismachined parts can be built

up with steels, copper and copper alloys, nickel and its alloys, or with ceramics.

OXYFUEL GAS POWDER SPRAYING

The surfacing material in the form of a powder is contained in a hopper from which it is conveyed into the spray flame either by reduced pressure generated by a venturi in the gas stream or in a carrier gas (usually air), Fig.3.6. Some designs of powder torch use additional air to propel the heated powder particles on to the substrate, to confine the spray stream, and/or to cool the substrate.

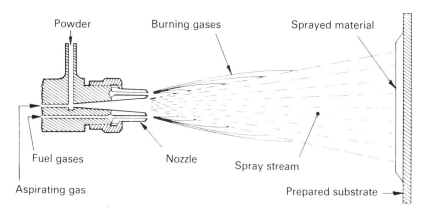

3.6 *Powder combustion spray method.*

Advantages
1 The most suitable flame spray process for high alloy and self-fluxing surfacing materials;
2 Gives a continuous supply of surfacing material;
3 Can be used to provide coatings of materials which cannot be produced as rods or wires.

Disadvantages
1 Requires consumables in powder form of suitable mesh range;
2 Gives lower density deposits of lower adhesion strength than arc spray (except in self-fluxing alloys subsequently fused).

Principal applications
Powder spraying is used extensively for application of wear resistant coatings of nickel or cobalt base alloys which are subsequently fused to give fully dense, metallurgically bonded deposits.

Metal powders such as steels, copper alloys, nickel alloys, and others are used to restore worn surfaces.

ARC WIRE SPRAYING

Electric arc wire spraying uses twin wires which are fed together into a spray head in which an arc is struck between the wire ends and the resulting molten material is projected from the arc as a series of molten droplets by means of a high pressure gas jet (usually air), Fig.3.7 and 3.8.

3.7 *A heavy duty Arcspray pistol for mechanised and automated plant rated at up to 600A (courtesy Metallisation Ltd).*

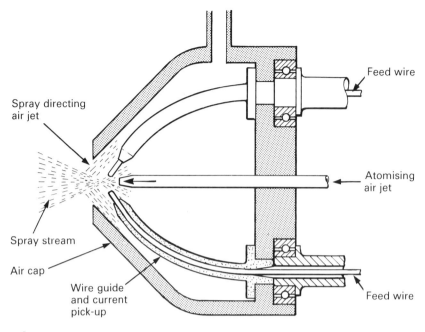

3.8 Electric arc spray torch.

Advantages
1 Either solid or tubular wires can be used giving a wide range of coating alloys;
2 Very high spray rates are possible;
3 By using two different wires composite or 'pseudo-alloy' coatings can be produced.

Disadvantages
1 Limited to consumables available in wire form;
2 Lower density deposits of lower adhesion than plasma, D gun and Jet Kote*.

Principal applications
The arc spray process is most suitable for surfacing large components such as rollers for the steel, paper, paint, and plastics industries, and to surface and reclaim hydraulic rams and pistons, shafts, bearings, *etc.*

PLASMA ARC SPRAYING IN AIR (APS)
A plasma arc spray torch consists of a tubular copper anode in the rear of which is a tungsten cathode; both electrodes are water cooled and are surrounded by an insulating body which holds them in correct relation to each other and serves as an arc chamber. A high current arc is generated within the torch and a gas injected into the arc chamber where it is heated and, on passing through a constriction in the anode bore, is converted into a high temperature plasma. Powdered surfacing material is injected into this plasma jet and is thus heated and accelerated on to the substrate, Fig.3.9 and 3.10.

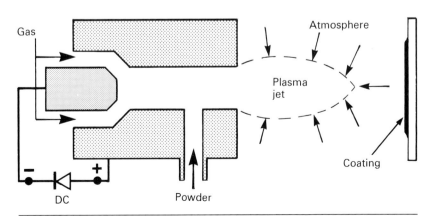

3.9 Plasma arc spraying in air.

*Jet Kote is a trademark of Stoody Deloro Stellite Inc.

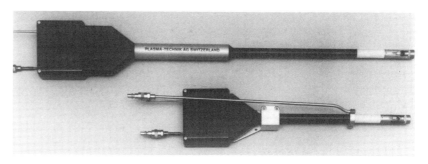

3.10 *Plasma spray torches have been developed in a variety of configurations to satisfy the requirements of particular work, involving manual and mechanised installations and for application on external and internal surfaces (courtesy Plasma-Technik AG).*

Advantages
1 The high temperature enables almost all materials to be sprayed;
2 Deposits are of high density and strongly bonded to the substrate;
3 Very low heat input to the substrate.

Disadvantages
1 Higher capital cost than gas and arc spraying, bulky equipment;
2 Deposits of lower density and adhesion than vacuum plasma spraying (VPS).

Principal applications
Used mostly for deposition of refractory, high melting point materials, ceramics, carbides, and high temperature alloys on to aircraft engine components, textile machine parts, wear resistant coatings on pump and valve spindles, and application of electrical and thermal insulating coatings.

VACUUM PLASMA SPRAYING

This process, known as vacuum plasma spraying or low pressure plasma spraying (VPS/LPPS), Fig.3.11, involves spraying in a chamber initially evacuated to a pressure of 10^{-2} mbar and then backfilled with argon to 40 mbar.

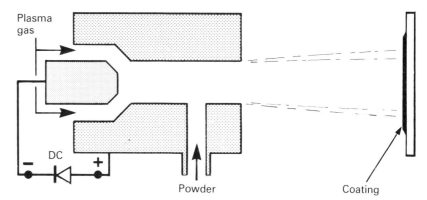

3.11 *VPS/LPPS process*

By excluding oxygen from the environment, neither the sprayed particles or the substrate surface becomes oxidised during spraying and deposits are dense and well bonded. The plasma jet is longer than in air and the workpiece reaches higher temperatures.

Advantages
1 High density, high adhesion, low oxide coatings;
2 Coating properties reproducible to close limits.

Disadvantages
1 High capital cost, bulky, fixed installation;
2 High heat input to workpiece may cause distortion or metallurgical changes in the substrate;
3 Not a manual process.

Principal applications
Figure 3.12 shows an installation incorporating a robot manipulator with a rotary table positioner used for coating developments. The high quality

3.12 *Developments in plasma spraying include use of robotics and low pressure plasma — the interior of the Materials Engineering Centre's chamber facility at Harwell (courtesy of AERE Harwell).*

coatings produced by this process have led to applications on aero-engine parts which are coated with materials such as the M-Cr-Al-Y series to resist corrosion and wear at high temperatures.

Figures 3.13 and 3.14 show the microstructures of 50-50 Ni-Cr alloy deposited by VPS/LPPS compared with APS and illustrate the signficant increase in density and reduction in oxide obtained by low pressure operation.

3.13 *Vacuum plasma sprayed deposit of 50-50 NiCr alloy (courtesy of Plasma Technik Ltd).*

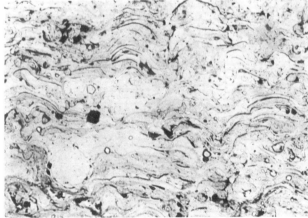

3.14 *Air plasma sprayed deposit of 50-50 NiCr alloy (courtesy of Plasma Technik Ltd).*

HIGH VELOCITY COMBUSTION SPRAYING

This group of processes is designed to give high levels of coating density and adhesion to the substrate and can handle coating materials (in powder form) of weldable and non-weldable types.

The density and adhesion of the coating depend on a number of factors which includes the temperature and the velocity of the sprayed particles when they strike the workpiece. Higher values of these factors are achievable from the specially designed spray guns, burning oxygen and a fuel gas, than those obtained from conventional gas and arc spraying processes.

The actual particle velocities in practice depend on the process, the gun, the gases used, the spraying parameters and the material being sprayed. Initial gas velocities and temperatures in the guns have been reported in the region of 1300 m/sec and 2875°C with particle velocities of 400-800 m/sec.

This is an interesting field of technology undergoing continuing development; information generated on applications for the newer processes will assist potential users in process and coating material selection.

Detonation coating

A device similar to a rifle barrel has powdered coating material and an oxygen/acetylene gas mixture metered into it, which is then ignited by a spark discharge, Fig.3.15 and 3.16. The mixture detonates and the powder is propelled from the barrel at high temperature and velocity. The operating cycle is repeated to provide a continuous deposit. Because of the noise created by the process it is carried out in a soundproof chamber and is fully automated.

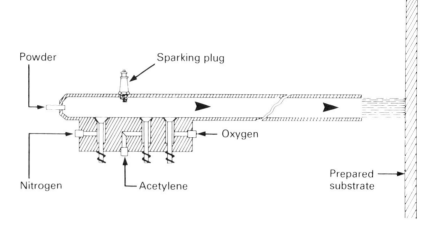

3.15 Detonation gun.

3.16 Detonation gun in operation on gate valve components (courtesy of Union Carbide Coatings Division).

Applications include protection of gas turbine engine parts against various forms of wear at high temperature, textile machine parts, rollers for steel, paper and plastics industries, bearings and other parts for the nuclear power industry; cutting edges.

Advantages
1 Very high density deposits;
2 High deposit adhesion;
3 Heavy grit blast surface preparation is not needed;
4 Low heat input to workpiece;
5 Fully mechanised and controlled process.

The Jet Kote process

The Jet Kote gun uses continuous combustion of oxygen and fuel gases to accelerate and project powdered coating materials on to a workpiece surface at high velocity to form dense and strongly adherent coatings, Fig.3.17. All materials are applied direct to the substrate, no bond coat being required.

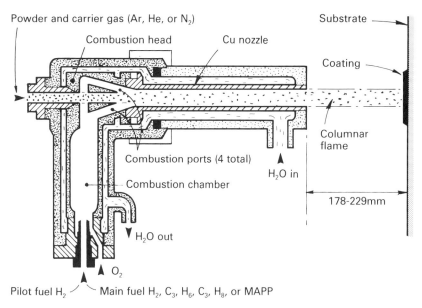

3.17 *Cross section of Jet Kote gun.*

The process can be used manually or in a mechanised and controlled set-up and either conventional or mass flow gas controls are available. Thus both fixed and on-site operation is possible.

The process has been developed for deposition of materials which include carbides of tungsten and chromium, and for alloys including stainless steels, cobalt and nickel base. Applications are already established or in the process of evaluation in the textile, pipe line valve, chemical, power generation and aircraft industries.

Figure 3.18 shows the equipment in use manually, while Fig.3.19 shows the high quality of deposit attainable, in this case, 83/17 tungsten carbide/cobalt on a titanium substrate.

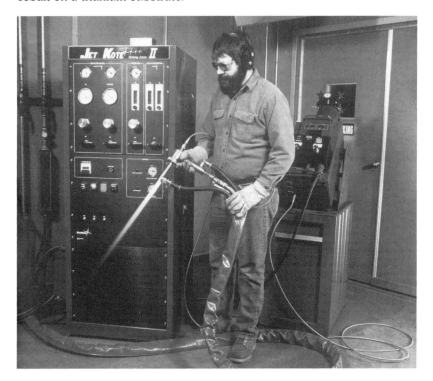

3.18 *Jet Kote equipment in manual form (courtesy Stoody Deloro Stellite Inc).*

Advantages
1 Deposits of high density and adhesion to the workpiece;
2 Low heat input to the component;
3 Tight spray pattern allows accurate placement of the deposit;
4 Gun-workpiece distance relatively insensitive;
5 Manual or mechanised capability.

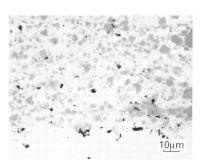

3.19 *Deposit of 83/17 WC/Co on titanium (courtesy BAJ Ltd, Coatings Division).*

The Diamond Jet system

The Diamond Jet system uses a controlled flow of oxygen and a fuel gas (typically propylene or hydrogen) together with compressed air to accelerate molten material from the gun to strike the workpiece at high velocity, Fig.3.20. It is designed to produce high quality sprayed coatings of metals, carbides and special materials such as abradable deposits for gas turbine engines.

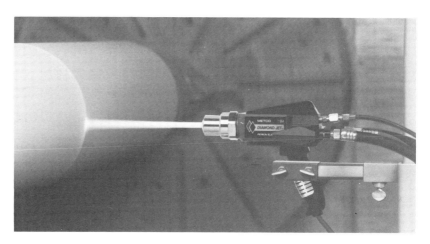

3.20 *Mechanised spraying of a roll surface using the Diamond Jet process (courtesy Metco Ltd).*

The process is available in manual, mechanised and computer controlled forms to suit a variety of industrial requirements. Air cooling of the gun eliminates the need for water cooling systems.

An interesting feature of the process is that coating thickness limits have been observed to be greater than those of a similar standard, sprayed by alternate processes. This is attributed to the excellent adhesive and cohesive bonding and to reduced stress within the coating.

Advantages
1 Dense high bond strength coatings;
2 Thick low stress coatings independent of coating hardness;
3 Torches available for spraying internal surfaces down to 100mm diameter;
4 Simple, easy to operate and maintain.

The CDS process

This is described as a continuous detonation process which uses oxygen with propane or propylene as the fuel gas and is designed to apply dense adherent coatings of excellent wear resistance, using materials such as tungsten carbide/cobalt, chromium carbide/nickel chrome, Tribaloy, Inconel and ceramics in powder form.

The process is supplied as a complete spraying system equipped with advanced gas controls. The gas combustion is designed to give significantly higher velocities which can be achieved with less consumption of fuel gas and oxygen. This is said to provide a broader range of useful operating parameters and more precise optimisation of the process compared with other supersonic processes.

Advantages
1 High reproducibility of set spray values;
2 Single or dual powder feed system;
3 Wide range of velocity control;
4 Dense coatings of high adhesion to the substrate.

Design for thermally sprayed coatings

As with weld deposited coatings, there are many factors to be considered when drawing up a design for thermally sprayed coatings which are discussed below.

COMPONENT SHAPE

Surfaces to be coated must be accessible to the spray guns available for the chosen process and the component must provide adequate routes for dispersion of the hot spray gases and overspray. It must be possible to direct the spray at right angles to the surface, although sometimes this may be reduced to 45° at the expense of increased overspray and some loss of adhesion to the substrate.

Thus re-entrant areas of any type may produce a problem which increases with the depth/width ratio (unless a gun designed to work inside the recess is available) and if there is no through passage providing adequate ventilation for gas and overspray.

If the recess performs no practical function, it may be easier to seal it off with a welded-in plug, which after grinding flush, leaves a smooth, clean external surface to be sprayed.

SURFACE PROFILE DETAILS

Fabrications

Figure 3.21 shows joint configurations which should be used and draws attention to the problems of unsealed surfaces and discontinuous fabrication welds. Unsealed joints can harbour dirt and moisture which can give rise to local defects in the sprayed coating and sites for premature failure.

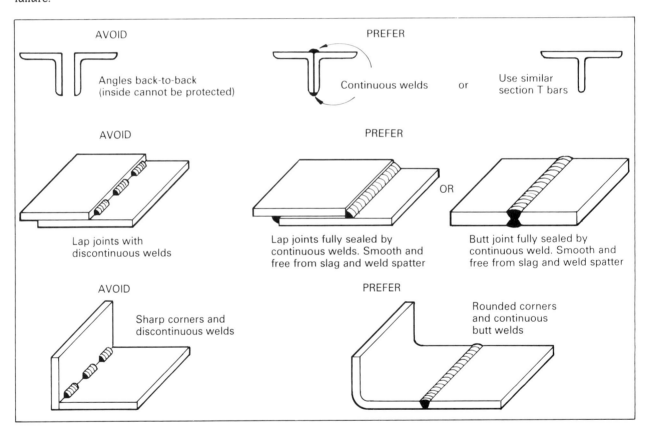

3.21 *Crevices, lap joints and corners.*

Precision components

If the whole surface is not to be covered it is normal to arrange for the edges of sprayed deposits to finish flush with the parent material of the workpiece. This involves creation of a recess to accommodate the required finish thickness of the deposit and the provision of a finishing allowance on adjacent areas. Bearing in mind the low heat input of most thermal spraying processes and the low risk of distortion, this finishing allowance can normally be quite small, typically less than 0.5mm per face. The important feature of the recess is that it should be provided with the edges chamfered full depth at an included angle with the base of the recess not more than 45°; 30° would be better, Fig.3.22.

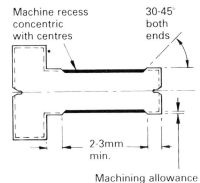

3.22 *Design features for a thermally sprayed shaft.*

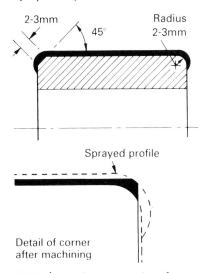

3.23 *Alternative preparations for corners of sprayed deposits.*

When deposits must reach the end of a shaft or bore, the corner should be provided with a radius or chamfer as shown in Fig.3.23 but a heavy local build-up of deposit (in an attempt to machine to a square corner) must be avoided. This would encourage additional stress and a possible risk of the deposit lifting.

Corners of deposits produced in the manner described are always weaker than a deposit contained in a recess and the latter is to be preferred especially if the part is likely to be exposed to shock on the corners when handling or in use.

In reclamation work there is always the possibility that holes or keyways will be present in the surface to be coated. The usual way to deal with these is to plug them with graphite, made a tight fit in the recess and protruding slightly higher than the finished thickness of the deposit. The plug will be revealed on machining to finished size and being soft is easily removed. However, this leaves a sharp and fragile edge on the deposit which should be radiused, chamfered or tapered to remove the sharp corner and to minimise damage in use.

DEPOSIT THICKNESS

Factors to be considered concerning different thickness vary for new and for reclamation work. For new work, the best guide is to apply the minimum necessary to do the job taking into account any opportunity for regrinding before recoating is necessary. For repair work, thicker deposits may be necessary. Excessive wear can often be made good by spraying a first coating of a softer material and then finishing with the desired top coating. This is because soft materials applied by low energy processes can be applied in thicker coatings having lower residual stress than hard materials, particularly if applied by high energy techniques. An indication of typical thickness design figures is given in Table 3.8.

In reclamation work, irregular wear patterns should be machined to avoid a varying deposit thickness, but local increases in thickness can be accommodated using a 30 or 45° chamfer at each change in depth.

Coating thickness for corrosion protection on iron and steel structures sprayed with zinc or aluminium ranges from about 100-400µm and 100-250µm respectively, the latter coatings being used for longer life structures.

Table 3.8 Typical design figures for thermal spraying

	Gas temperature, °C	Particle velocity, m/sec	Adhesion, MPa	Oxide content, %	Porosity, %	Spray rate, kg/hr	Relative cost, low = 1	Typical deposit thickness, mm
Flame	3000	40	8	10-15	10-15	2-6	1	0.1-15
Arc wire	N/A	100	12	10-20	10	12	2	0.1 to > 50
HVC Jet Kote	3000	800	> 70	1-5	1-2	2-4	3	0.1 to > 2
D gun	4000	800	> 70	1-5	1-2	0.5	N/A	0.05-0.3
APS	12000	200-400	4 to > 70	1-3	1-5	4-9	4	0.1-1
VPS	12000	400-600	> 70	ppm	< 0.5	4-9	5	0.1-1

Coating production

SURFACE PREPARATION

The importance of correct surface preparation for all thermally sprayed coatings has already been stressed as one of the main factors in achieving a high and consistent strength of bond between coating and substrate.

The first requirement is removal of any grease and dirt, if possible by a degreasing operation. The second is roughening of the surface to promote an effective key with the sprayed particles. This can be achieved most readily by abrasive blasting, although for metallising processes rough machining (e.g. by thread cutting) has been used effectively.

It is important that a blast medium is kept clean, free from fines and, for metallising, an angular chilled iron grit is used on most substrates and this must be replaced when it becomes blunt. Grit sizes range from 14-100, see Table 3.9.

Table 3.9 Grit sizes used for surface preparation

Grit size	BS 2451 equivalent	Nominal sieve opening, mm	Size distribution, mm
14	G55	1.41	2.0-0.84
18	G39	1.0	1.68-0.71
25		0.71	1.0-0.42
40	G17	0.42	0.59-0.25
60		0.25	0.35-0.15
100		0.15	0.18-0.06

Smaller size particles permit blasting of a larger area per hour. Larger particles provide more rapid removal of material and give a rougher surface finish. Grit sizes 16 and 60 are used for metal substrates and 60-100 for most plastics. For thin coatings, particularly when used on thin substrates, fine grit should be specified. Coarser grit (14-25) providing rougher finishes, is used for thick coatings, (greater than 0.25mm) and optimum coating adhesion.

On cylindrical components, the operation to machine a recess for the deposit is sometimes followed by cutting a V thread in the bottom of the recess, at say 0.5mm pitch, as this gives a good key to the sprayed deposit. This may be followed by abrasive blasting. All traces of cutting fluid must be removed. For coatings applied by high energy processes, aluminous grits in the 25-100 range are used.

The action of abrasive blasting increases the surface area of the workpiece significantly, and in its clean and abraded state, it is very active and oxidises readily. It should therefore be sprayed as soon as possible after the blasting operation and never handled directly by the fingers, which can transfer dirt and grease to the surface. Any contamination can impair adhesion of the coating.

The abraded area should always stretch beyond the area to be coated to prevent risk of the overspray lifting and the bond failure propagating under the coated area, even for a short distance.

On high quality work, such as aircraft turbine engine parts, automatic grit blasting machines have been developed which control air pressure, blast angles, nozzle standoff and time of blasting. Further quality assurance standards are used to control grit size range, nozzle wear and grit damage with a particular substrate.

Abrasive blasting can lead to distortion of thin components and a reduction in grit size or impact velocity may not always overcome this. Some materials to be coated may be too hard to roughen adequately by grit blasting and to overcome the risk that a coating may not adhere properly under these circumstances, a preliminary layer of a bond coat is applied.

BOND COATS

Some sprayed materials adhere strongly to clean smooth surfaces. They are adherent over a wide range of conditions and a thin layer can serve as an 'anchor' for materials sprayed on top. They are used in both of the circumstances mentioned in the paragraph above.

Commercially available bonding materials include nickel-aluminium and molybdenum.

Nickel-aluminium is used for a large number of applications because of its better high temperature properties and less dependence on operator technique. It is used on a wide range of substrates including all common steels, stainless steel, hardened alloy steel, nitriding steel*, cast steel, nickel, chrome-nickel alloys, Monel, cast iron, magnesium, aluminium, titanium and Kovar.

Coating thickness is generally specified at 75-125µm. It is sometimes advisable to preheat the workpiece to avoid any condensation on the cold metal surface; 90-95°C is adequate.

Molybdenum is widely used as a bond coat on all common steels, stainless steels, chrome-Ni alloys, cast steels, magnesium, most Al steels, Monel,

---

*Nitrided and carburised surfaces are best removed before spraying.

nickel and cast iron but does not bond properly to copper, brass, bronze, chrome plate, nitrided surfaces or electroless nickel.

Coating thickness is normally 30-80μm. The work must be preheated to a surface temperature of about 90°C to avoid condensation and further heating to 175-260°C improves the bond.

MASKING

With thermal spraying processes, use of a mask or shield prevents deposition on undesired surfaces. The mask should be positioned above the surface to be sprayed; if placed directly on the surface the coating will link the job and the mask; removal is then likely to damage the coating. Masks are conveniently made from sheet metal, cut or bent to the required shape and supported rigidly so that the force of the spray does not dislodge them.

SEALING

It is advisable to seal all deposits used in machine parts except those used in lubricated conditions under light load. This prevents ingress of fluids which

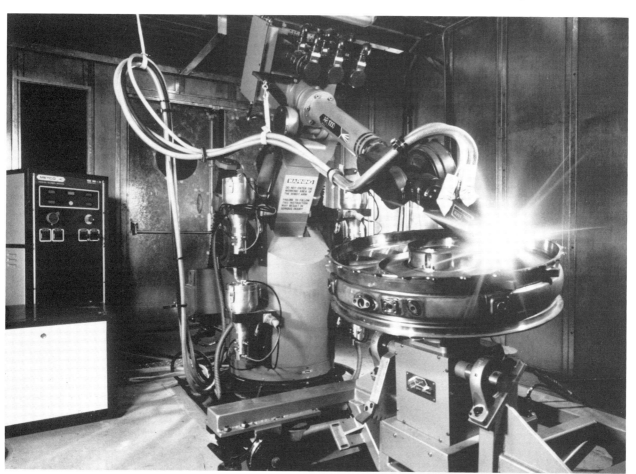

3.24 *A programmable 6-axis robot teamed with a 5-axis positioner can accurately reproduce preset spray patterns. A memory holds up to 10 pre-programmed spray patterns (courtesy Metco Ltd).*

3.25 a) *Deposit of flame sprayed 13% chromium steel;* b) *Arc sprayed deposit of the same steel using correct conditions;* c) *As for b but poor interface because of excessively fine atomisation and use of inadequate extraction (courtesy Metallisation Ltd).*

may accelerate corrosion of the coating or penetrate to the substrate, causing corrosion at the interface and causing the coating to lift. Surface porosity may be impregnated by materials being processed which could contaminate subsequent batches.

One of the most important reasons for sealing the coating on shafts to be used for journals, pump plungers, *etc*, is to ensure a better, cleaner ground finish. Grit from the grinding operation is prevented from entering the pores in the coating.

Sealers are low viscosity fluids containing an inert resin binder in a solvent. This impregnates the pores in the coating, filling them as the solvent evaporates. When used on coatings such as zinc, use of a sealer does not impair the cathodic protection of the coating. Silicone resins are available for service up to 550°C.

SPRAYING
Thermal spraying processes generate overspray which can constitute a health and safety risk. This subject is dealt with in detail in Chapter 9.

Emphasis has been placed on the importance of correct design and surface preparation of the component if satisfactory results are to be obtained. Equally important to the result is control of spraying conditions.

The ability of a correctly designed and selected deposit to give satisfactory service depends on such factors as coating density, oxide content and bond strength with the substrate. These are properties which are not easily verified by non-destructive testing, so when developing new applications, it is usual to carry out destructive tests on prototype or simulated components. When necessary adjustments can then be made to surfacing parameters until acceptable results are obtained.

The important next step is to ensure that conditions pertaining to the acceptable testpieces are accurately repeated on production work, and this means use of properly calibrated devices to measure each parameter. A mechanised process incorporating microprocessor control of parameters and, when appropriate, robotisation, Fig.3.24, will give improved repeatability compared with manual work and will always be the preferred choice for critical work.

The photomicrographs in Fig.3.25 illustrate structures of deposits of 13% chromium steel sprayed by flame and electric arc processes. The laminar structure characteristic of such deposits is clearly seen, with the arc deposit having lower porosity. By contrast, Fig.3.25c shows a deposit applied with unsuitable conditions; this would not necessarily be obvious on visual examination of the surface.

Spray fused coatings

The two stage spray fuse process was developed to improve the strength of bond between coating and substrate which is obtained with a normal thermal spraying process and at the same time take advantage of the typical smoothness and uniformity of coating (which can be quite thin, e.g. less than 1mm).

After applying the coating typically by oxyacetylene spraying and observing design requirements, a fusing operation follows which involves heating the coating and the underlying substrate to a temperature between the solidus and liquidus of the coating alloy, usually about 1000°C. This develops a metallurgical structure in the coating and bond with the substrate, which in mild steel has a strength of 380-550MPa, depending on coating alloy. To be successful the coating alloy requires self-fluxing characteristics and these are possessed by a range of cobalt and nickel base alloys which contain a proportion of boron and silicon. Self-fluxing alloys are also applied by a process known as powder welding which is dealt with in Chapter 2.

The spray fuse process lends itself particularly well to coating of cylindrical parts such as pump shafts and sleeves, for which spraying can easily be mechanised to give a uniform, smooth coating.

Fusing of the coating can be carried out in various ways. Manual fusing using an oxyacetylene torch fitted with a large multi-jet nozzle as shown in Fig.3.26

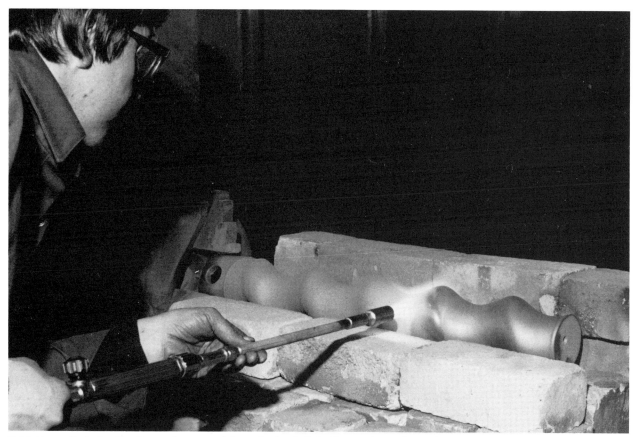

3.26 *Manually fusing the thermally sprayed deposit of a Cr-Si-B alloy on the surface of a pump rotor (courtesy Deloro Stellite Ltd).*

is widely practised. The large flame introduces heat gently into the component and this is important to avoid the coating heating too rapidly and expanding away from the substrate before it is fused. This would prevent attainment of a metallurgical bond when fusing temperature is reached.

A cylindrical component is rotated in, for example, a lathe chuck and the flame is directed at one end of the coating. At a temperature between the solidus and the liquidus a band on the surface of the coating develops a glossy, light reflecting finish, which indicates that the correct temperature has been reached. The torch is then traversed slowly across the coated area to produce this effect all over and the component is then allowed to cool, or is heat treated if required.

As the temperature in the component is relatively high, precautions similar to those used in weld surfacing must be taken to suit the substrate material, remembering that the heat affected zone is likely to be deeper than that associated with welding.

An alternative method of fusing involves transfer of the sprayed component to a vacuum furnace. With the heating, fusing and cooling cycle carried out in vacuum, the oxides formed during spraying are reduced and no further oxidation occurs. Sufficient time at fusion temperature is allowed for diffusion processes to take place within the coating and at the coating-substrate interface and this provides cohesion and adhesion strengths much superior to normal thermal spraying and close to those of welded deposits. With this process, the cooling stage can often be adjusted to suit the desired metallurgical structure in the substrate.

Vacuum fusing offers advantages over manual fusing shared with the spray fuse process. It:

— Is ideal for quantity production, using batch fusing techniques;
— Is easily controlled and repeatable giving consistent results;
— Provides uniform heating of the whole component imposing less stress on the bond between the coating and the substrate as the temperature rises;

- Avoids the problem of temperature control experienced when fusing coatings manually on parts whose cross section suffers sudden local changes;
- Enables components of thicker cross section to be coated than is practicable with manual torch fusing.

Figure 3.27 shows taper plug valves coated with an NiCrSiB alloy with a hardness of ~60RC. The coin of 22mm diameter gives a guide to the size of the plugs and the smooth, even, as-fused surface is clearly seen. On a component such as a ball valve, manual fusing would be difficult and incur the risk of the sprayed coating lifting from the substrate while other areas are being fused. Uniform heating in a vacuum furnace, Fig.3.28, avoids the thermal stresses which cause this problem and also minimises any tendency for the part itself to distort. These advantages of the spray and vacuum fuse processes mean that less material, time and machining is required than if a weld surfacing process were used.

3.27 Taper valve plugs thermally sprayed and vacuum furnace fused (courtesy Pro-Vacuum Ltd and Serck-Audco Ltd).

The temperature needed to fuse the coating can also be developed by induction heating, provided that the rate of heating and the uniformity of temperature required for satisfactory results can be achieved. A further possibility would be to use a laser beam, but this has so far seen little industrial application.

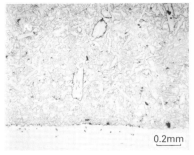

3.28 Sprayed deposit of Co-Cr-W-Ni-Si-B alloy, vacuum furnace fused (courtesy Deloro Stellite Ltd).

SUBSTRATE MATERIALS

The thermal cycle

It has already been noted that temperatures required to fuse sprayed deposits of self-fluxing alloys are in the region of 1000°C, and with manual fusing the whole part may exceed 800°C. With furnace fusing, the whole part will be at fusing temperature.

Information given in Chapter 2 on weld deposited coatings therefore applies to a large extent to fusing of a sprayed coating. Differences in treatment are discussed below.

After fusing has commenced, if a hardenable substrate steel is in use, no area of the fused component should be allowed to cool below the temperature at which martensite forms until the fusing operation has been completed and and the part subjected to a suitable heat treatment — as for welded deposits.

With large areas of coating and deposits of hard alloys, or on components of large mass, it is helpful to introduce extra heat into the component during spraying, once the whole area has received an initial thin coating of spray. On conclusion of spraying, and with a substrate temperature of ~500-700°C, there is less thermal shock and differential expansion stress on the coating when raising it to fusing temperature so risk of the bond to the substrate failing is minimised. Cooling stresses are also lower when fusing is carried out with less steep temperature gradients in the component. The first sprayed coating protects the substrate from oxidising as the temperature rises later in the spraying operation.

Unsuitable substrate materials

The component material must have a solidus temperature greater than the fusing temperature of the coating alloy. The presence of, for example, a high phosphorus content in a cast iron can result in melting of low melting point constituents at the grain boundaries and collapse of the part when fusing the coating.

Carburised and nitrided surfaces should be removed before surfacing and free-cutting materials should be avoided. Materials capable of forming tenacious oxide films caused, for example, by an aluminium content, can resist the self-fluxing action of the coating alloys and inhibit the formation of a strong bond which can result in detachment of the coating.

COATING ALLOYS

The alloys used for the spray fuse process are those of nickel base (Group 2 type 5) and cobalt base (Group 3 type 6), see Table 0.1. A wide range of hardnesses is available from ~ 20-60RC. Those of medium to high chromium content have resistance to oxidation and corrosion at elevated temperature and their abrasive wear resistance increases with hardness. Cobalt base alloys have good high temperature hardness retention, resistance to adhesive wear and erosion.

All the alloys are supplied in powder form for spray fusing and are frequently mixed with tungsten carbide in proportions ranging up to about 75%, to provide even greater resistance to abrasive wear in use.

COMPONENT DESIGN FOR SPRAY FUSE COATINGS

Deposit thickness

As with other processes, the thickness of the finished deposit is based on the amount of wear the surface can tolerate. On most jobs, 0.5mm should be the minimum design figure after grinding to size. Maximum thickness is related to the method used for fusing the deposit and the need to ensure that the interface with the substrate reaches the correct fusing temperature without the surface exceeding the melting point.

To the finished thickness of deposit must be added allowances to cover possible distortion arising during the coating or subsequent heat treatment operations, and a small machining allowance if the part is to be precision machined.

The size of these allowances depends on the design and size of the part, particularly matters like length/diameter ratio, which influence distortion. On short, stubby shafts which have been accurately centred at the ends and can therefore be accurately set up for machining, a combined allowance of 0.1-0.5mm may be sufficient if spraying is effected mechanically to ensure uniformity.

An extra allowance must be added to arrive at the size of the part in the as-sprayed condition before fusing. This is to compensate for two factors: the first is that the sprayed coating shrinks some 12-20% on fusing, depending on the alloy and spraying conditions. The second is that, when measured, the sprayed coating and the component will be hot and therefore will have expanded.

These various allowances can readily be identified on production repetition work and an optimum condition established for future work. In jobbing work, it may be cheaper to be a little generous with the machining allowance rather than run the risk of rework.

Surface profiles

The designs for parts to be thermally sprayed described earlier are equally applicable to spray fusing.

SURFACE PREPARATION BEFORE SPRAYING

As with thermal spraying, the surface of the component is roughened by blasting with angular iron grit, typically BS 2451 G39 grade immediately before spraying (see earlier and Table 3.9 but note that bond coats are not applicable to the spray fuse process).

PREHEATING

Preheating the prepared surface to prevent condensation of water vapour is normal, good shop practice. A temperature of 90°C is adequate. This operation should follow as quickly as possible after shot blasting to avoid risk of the surface becoming contaminated with rust or dirt.

On cylindrical parts, preheating should be carried out while held in a rotating chuck so that heat intake is as uniform as possible.

PRACTICAL TECHNIQUE

Spraying

Spraying should be as near normal to the surface as possible and never at an angle of less than 45°, so adequate access is essential. Special spray guns are available which enable bores of 150mm diameter to be sprayed to a depth of 500-600mm from an open end.

The spraying operation should be mechanised where possible to ensure a uniform coating thickness and economy in time and material.

As already discussed, areas which are to be kept free from deposit can be masked with suitable tape to prevent damage during blasting and a metallic shield can be used as a shadow mask to limit the edges of the deposit. This must stand clear of the surface so that its removal does no damage to the deposit.

Manual fusing

As indicated earlier, to ensure the creation of a metallurgical bond between the sprayed particles and between the coating and substrate, the fusing temperature for the coating alloy must be reached throughout its thickness. This is achieved when temperature gradients are as low as possible; a condition readily achieved by furnace fusing but requiring some care when torch fusing.

The most important point to observe is that heat is introduced gradually as it has to be transmitted through the thickness of the coating into the substrate. Too great a rate of heat input can result in the coating expanding away from the substrate, a condition which requires the damaged coating to be removed and the job restarted.

Once a thin coating has been applied by spraying and the substrate is thereby protected against oxidation, it is good practice when torch fusing is intended, to use extra heat if necessary to ensure that the whole of the sprayed area of the component rises in temperature as spraying proceeds. When this stage is reached and the component is at perhaps 600°C, extra heat can be applied generally with the fusing torch before this is concentrated at one end of the deposit and fusing is commenced. This technique ensures that temperature gradients are kept low and satisfactory coatings are properly bonded to the substrate.

Figure 3.29 illustrates a nickel base alloy coating which requires a temperature of 1000°C minimum at the interface. With too low a temperature in the substrate, the coating may be fused but not bonded to the substrate. In fusing by induction heating, the substrate actually exceeds the coating in temperature, even when the core temperature is low.

Conclusions

Thermal spraying processes provide a wide range of possibilities for application of coatings which are capable of giving high resistance to wear, corrosion, oxidation and heat as well as those which can be used for thermal or electrical insulation.

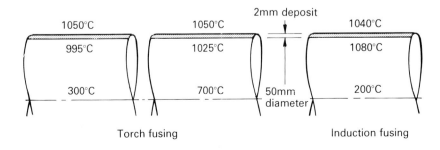

3.29 Temperature differences in manual fusing of a nickel base alloy coating.

Production of reliable coatings depends on:
— Selection of the correct coating material;
— Use of the correct coating process;
— Proper surface preparation;
— Adherence to correct operating parameters;
— Process automation;
— Precise quality control of powder consumables.

Electrodeposited coatings

CHAPTER 4

Electrodeposition is a well established process for applying metallic coatings to improve surface properties of materials used in engineering practice. Although the principles are similar to those involved in application of relatively thin coatings for decorative and corrosion protection purposes, engineering electrodeposition differs in several important respects and constitutes a specialised process in its own right. For the sake of completeness, this chapter has been extended to describe application of certain metal based coatings by chemical reduction and by anodic oxidation. Both processes operate in aqueous environments and are often used in conjunction with electroplating processes.

In theory, there is no limit to the thickness to which many metals and alloys can be electrodeposited, but the thickness needed to perform the required function is usually the basic criterion. Process economics are always important, therefore it should be borne in mind that electrodeposition can be slow and costly, so where thick coatings are desired other surfacing methods may be more cost effective. For thinner engineering coatings, however, electrodeposition is essential to the successful operation of innumerable components, and it offers considerable scope and flexibility to the designer.

Basic principles

Electrodeposition or electroplating involves making the component to be coated the negative electrode or cathode in a cell containing a liquid which must allow the passage of electric current, Fig.4.1. This liquid or electrolyte, which is usually a solution in water of a salt of the metal to be deposited, is maintained at a temperature up to about 60°C. The electrical circuit is completed by a positive electrode or anode which is generally made of the metal to be deposited and is located a short distance away from the cathode. Under the action of a direct current applied at a low voltage, positively charged metal ions in the electrolyte move towards the cathode where they undergo conversion to metal atoms and deposit on the cathode, *i.e.* the component surface.

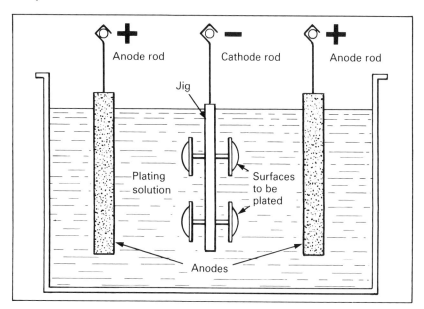

4.1 *Plating bath.*

The structure and properties of the deposited metal depend upon the chemical composition of the electrolyte as well as its temperature and degree of acidity or alkalinity, *i.e.* its pH value. These factors, and in particular the density of the electric current per unit area of cathode surface, determine the rate of deposition. Current flow to projections on the cathode surface is greater than to recesses and the consequent variation in current density influences metal distribution as the thickness deposited is proportional to the current density. Uniformity of thickness, which is a function of the throwing power of the electrolyte, can be improved considerably by attention to the design of the component and to the conditions of deposition. As electrodeposited coatings are seldom of constant thickness over the entire surface, it is usual to define that portion of the surface which is essential to the serviceability of the component as the 'significant surface' and to quote the minimum rather than the average thickness over this surface.

Electrodeposition is used extensively not only to apply coatings to new components to confer the required surface properties, but also to restore the dimensions of parts which have either worn excessively in service or been so over-machined as to be outside required tolerances. Another application is production of free-standing bodies by deposition on to shaped mandrels which are capable of subsequent removal. This process, known as electroforming, enables the shape and surface finish of the mandrel to be faithfully reproduced and in this way constitutes a simple method of fabricating parts of intricate shape in a single operation.

Two processes are available for electrodeposition of coatings for engineering purposes, following the principles outlined earlier. The characteristics of both processes and the complementary processes of electroless or autocatalytic deposition and hard anodising are described below.

VAT PLATING

Principles
Virtually all electrodeposition for engineering applications is undertaken in tanks or vats which may have capacities of up to several thousand litres. For some specialised applications, the tank may be built round the workpiece or a large cylindrical component requiring plating internally may function as its own tank. The electrolyte is usually mildly agitated either by air jets or by mechanical movement of the work and is maintained at the working temperature by electric immersion heaters or steam coils, alternatively, water cooling may be necessary. The anodes, suspended some centimetres from the workpiece, are sometimes inert in that they carry the current but do not dissolve in the electrolytic process, so the coating material is derived wholly from the metal in solution. The workpiece itself is mounted on a rack or jig and suspended in the electrolyte. A transformer-rectifier set normally supplies the plating current at a voltage in the range 4-8V. The current applied is read on an ammeter and the time to deposit the required thickness is estimated from this current and the known efficiency of deposition.

Deposition efficiency is taken as the ratio of the weight of metal actually deposited against the weight that should have been deposited by the current-time combination used (some of the current passed is usually wasted in the unavoidable evolution of hydrogen at the cathode as described later in this chapter). Deposition efficiency varies with the electrolyte system in use, the plating conditions employed and current density.

Non-metallic particles may be incorporated into the metal deposit, for example silicon carbide in nickel, by maintaining the particles in suspension in the electrolyte. The process is controlled by measuring the density and acidity of the electrolyte and, in the longer term, by chemical analysis.

A variation in vat plating enables small components to be plated in bulk. For example, components may be held in a perforated cylindrical barrel constructed in a plastics material which is immersed in the plating solution and rotated continuously. The work in the barrel forms the negative electrode and in tumbling over each other the components maintain electrical contact and simultaneously present fresh surfaces to the action of the electrolyte. Barrel plating is restricted to components weighing less than about 500g which are of simple shape and capable of tumbling without locking together. Coatings so applied are thin and the main applications are for decoration but engineering applications include deposition of gold and platinum group metals for corrosion and wear resistance.

Characteristics

1. As operating temperatures never exceed 100°C, the work should not undergo distortion or undesirable metallurgical changes.
2. Plating conditions may be adjusted to modify hardness, internal stress and metallurgical characteristics of the deposits.
3. Coatings are dense and adherent to the substrate. Bonding, which is molecular in nature, may be as strong as 1000 N/mm^2.
4. The thickness of deposit is proportional to current density and length of time of deposition.
5. As the current density over the workpiece surface is seldom uniform, coatings tend to be thicker at edges and corners and thinner in recesses and at the centre of large flat areas.
6. The rate of deposition seldom exceeds 75μm/hr, but it can be accelerated by forced circulation of the electrolyte.
7. There is no technical limit to the thickness of electrodeposits. Metals such as nickel may be 13mm or more in electroforming and reclamation work but most surfacing applications require much thinner coatings.
8. Application of coatings is not confined to the line of sight. Although the throwing power (*i.e.* the ability to plate round corners) may be limited, there is comparative freedom for the location of anodes, for example, in coating the bores of narrow tubing.
9. Areas not requiring deposition can be masked.
10. The size of the vat limits the dimensions of the work.
11. The process is suitable for automation.

BRUSH OR SELECTIVE PLATING

Electrochemical reactions in brush or selective plating are the same as in vat plating and the essential features are shown in Fig.4.2. The main distinction is that coatings can be manually applied to components such as moulds, dies, shafts and bearings in selected areas.

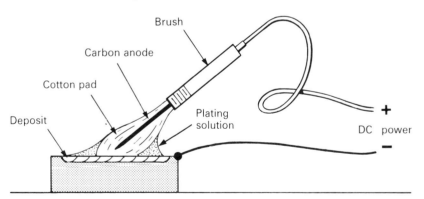

4.2 *Brush or selective plating.*

Principles

This process complements vat plating in that electrodeposits can be produced on localised areas with minimum masking or stopping-off and without immersion of components to be coated. Principles are the same as for vat plating, but the anode is mounted in an insulated handle and envelope by an absorbent pad which is soaked in the electrolyte. The work is connected to the negative side of a DC source and upon bringing the pad into contact with the work, the circuit is completed and electrodeposition takes place. Generally the operator traverses the pad manually over the areas to be coated but this procedure required much skill as the rate of movement is critical in ensuring a sound and even deposit (see later).

The anode is a high grade carbon rod and so does not dissolve. Long-fibre cotton wool may form the absorbent pad, but a resin-bonded plastics felt is preferable. The electrolyte may be renewed on the pad by dipping into a beaker, by drip feed or by pumping. Cleaning of the work before plating involves reversing polarity to make the work anodic in the presence of special solutions.

Characteristics

The first five characteristics of vat plating also apply to selective or brush plating. Additional characteristics include:

1 The equipment is portable and can be taken to the work, so that moulds and dies, for example, can be plated in situ with minimum loss of production time;
2 Masking is generally unnecessary, as only the area to be coated needs be in contact with the electrolyte;
3 The electrolyte must be more robust than in vat plating to withstand inevitable sharp changes in operating temperature and current density;
4 The power source, which operates in the range 8-30V, incorporates cut-outs to minimise damage by short circuits and also a current-time meter to monitor the progress of plating;
5 Deposition rates generally exceed those of vat plating and may reach 200-400 µm/hr;
6 The electrolytes are costly, but small volumes only are generally required;
7 Manual operation is labour intensive and demands considerable skill;
8 The process can be mechanised or automated, for example in plating cylindrical bores;
9 Brush plating is not suited to high volume production.

ELECTROLESS OR AUTOCATALYTIC DEPOSITION

Principles

Electrodeposition involves the reduction at the cathode (*i.e.* workpiece) surface of metal ions arriving from the electrolyte to produce metal atoms which are deposited. The current required for this reduction need not be supplied externally; copper, for example, can be plated on to iron by displacement. The coatings are very thin and, often, so poorly adherent as to be generally of limited use industrially. However, some metals such as nickel, copper and palladium can be deposited from aqueous solutions of their salts by chemical reduction in which the initial layer catalyses the subsequent deposition. Hard nickel alloys, deposited from solutions containing either phosphorus or boron compounds as reducing agents, are widely used in engineering. The process is generally operated in polypropylene or PTFE-coated stainless steel tanks containing the solution maintained at about 90°C, and fitted with facilities for accurate temperature control, agitation and solution filtration. The work must be cleaned just as efficiently as in electrodeposition.

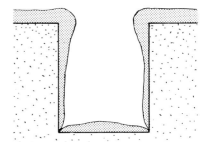

Electroplated nickel

Electroless nickel

4.3 *Comparison of metal distribution by electrodeposition and electroless deposition.*

Characteristics

1 The equipment is simple and economical as neither anodes nor DC electrical sources are required.
2 Deposits are uniform in thickness provided that the solution can be circulated over all the surfaces to be coated, Fig.4.3.
3 The rate of deposition, which is dependent on temperature, is approximately 20 µm/hr.
4 Thicknesses of up to 125-200µm may be applied.
5 Areas not requiring deposition can be masked.
6 The size of the tank limits the dimensions of the work.
7 Although the chemical materials are expensive, costs may be competitive with electrodeposition, especially when only a few components are processed.

HARD ANODISING

Principles

The objective of hard anodising is to thicken and strengthen the naturally formed thin tenacious oxide film normally present on metals such as aluminium and titanium. Hard anodising of aluminium and its alloys, which is a well established commercial process, involves making the work anodic in a tank containing an acidic electrolyte, typically 10-20% sulphuric acid. Upon applying a direct current at a sufficiently high potential, oxygen generated at the anode oxidises the aluminium and a balance is established between this action and dissolution in the electrolyte. Operation at low temperatures, *i.e.* 0-5°C, and high current densities combined with solution agitation favours oxide formation while suppressing the dissolution effect. The coating has a high resistance and so the starting voltage of about 25V often needs to be raised progressively to about 70V. In some anodising processes, alternating current is superimposed on the direct current to improve the wear resistance of the coating.

Characteristics
1 The equipment is more involved than that in electrodeposition as a refrigeration unit is often required to maintain the required temperature.
2 Hard anodising is a conversion process, *i.e.* the coating grows from the metal and is integral with it.
3 Throwing power, *i.e.* uniformity of coating thickness, is excellent.
4 Hard anodic coatings can be applied to aluminium and its alloys by selective or brush techniques.
5 Inserts of 'foreign' metals, *e.g.* steel, must be masked to avoid dissolution.
6 The process can be automated.

COATING MATERIALS
Electrochemical principles govern the mechanism of electrodeposition. Standard electrode potentials, recorded in Table 4.1, represent the electrical potential established between the metal and a solution of one of its salts under standard conditions, and they indicate the relative ease of deposition

Table 4.1 Electrode potentials of engineering metals

Metal	Standard electrode potential at 25°C, V (versus hydrogen electrode)
Aluminium	−1.66
Titanium	−1.63
Manganese	−1.18
Chromium	−0.74
Iron	−0.44
Cobalt	−0.28
Nickel	−0.25
Tin	−0.14
Lead	−0.13
(Hydrogen)	(0)
Copper	+0.34
Silver	+0.80
Gold	+1.68

of individual metals from aqueous solutions. In practice, the actual potential of the electrodes depends on the ambient conditions, in particular the solution concentration. Binary and ternary alloys may be electrodeposited provided that the plating conditions are chosen to equalise the deposition potentials of the constituent metals. Metals with higher negative potentials than −1.2V (Table 4.1) generally cannot be deposited from aqueous solutions, thus aluminium and titanium can only be plated either from solutions in organic solvents or from molten salt electrolytes, both of which necessitate specialised equipment and techniques. Nevertheless these processes are becoming available commercially.

Table 4.2 summarises the mechanical and physical properties of the metals and alloys more usually electrodeposited for engineering purposes.

Table 4.2 Mechanical and physical properties of some electroplated and electroless metals deposited for engineering purposes

Coating	Usual finished thickness, mm	Melting point, °C	Density, kg/m³	Hardness, HV	Tensile strength, N/mm²	Elongation, %	Resistivity at 20°C, μΩcm	Thermal conductivity, W/mk	Linear coefficient of expansion, ×10⁻⁶ per °C
Chromium	Up to 0.13 Sometimes to 0.50	1878	6930	800-1000	100-550	<0.1	14-67	67	7.4
Electrodeposited nickel	Up to 0.50 and even 13+	1455	8907	150-450	340-1050	3-30	7.4-11.5	94	13.5
Electroless nickel — 9% phosphorus	0.013 to 0.05	890 (approx)	8000	480 (1030*)	758 (69*)	1-1.5 (0.2*)	60-80 (20*)	—	13.0-14.5
Copper	Up to 0.50 and even 13+	1084	8933	60-150	180-650	Up to 40+	1.7-4.6	403	16.5
65/35 tin-nickel	0.01 to 0.02	800+	8400	650-700	—	—	—	—	—
Silver	Up to 1.3	961	10500	60-120	240-340	12-19	1.6-1.9	427 (approx)	21.0 (approx)

* After heat treatment. Otherwise data refer to as-deposited metals.
— Data either not available or not known accurately.

Chromium
Although its hardness when electrodeposited can be matched by other surface treatments, chromium possesses a unique combination of properties of value in engineering practice. The aggressive nature of the electrolytes makes the deposition of chromium alloys difficult, but claims for a 1% molybdenum alloy include improved resistance to both mechanical wear and corrosion in acid environments.

UK specifications for engineering chromium plating include BS 4641, DEF STAN 03-14/1 and DEF GUIDE DG13.

Characteristics
1 High hardness, i.e. 800-1000HV conferring resistance to abrasion.
2 Low frictional coefficient and resistance to sticking thus combatting adhesive wear.
3 Resistant to corrosion, also to oxidation up to 800°C.
4 Retains room temperature strength up to about 300°C.
5 Deposits thicker than about 50μm require finishing by grinding.
6 Thickness generally limited to about 0.5mm, but thicker deposits for reclamation work are usually built up on an undercoat of nickel.
7 Brittle, not resistant to shock loading.
8 Tensile stresses are sufficiently high to induce cracking and so coating thickness should not be less than 50μm for corrosion protection.
9 The crack pattern may be developed to produce an open porous structure for lubricant retention.
10 Deposition efficiency of chromium is low and so the process is energy intensive.
11 The concomitantly discharged hydrogen may dissolve into the workpiece.

Applications
1 Plastics moulds.
2 Metal forming and drawing dies.
3 Cutting tools.
4 Gauges.
5 Cylinder liners.
6 Piston rings.

7 Crankshafts.
8 Hydraulic rams.
9 Reclamation of worn parts.

Nickel

Nickel is used extensively in engineering to combat mechanical wear, corrosion, fretting and heat scaling. The range of plating solutions is wider than for chromium, and nickel is more economical to deposit. BS 4758 and DTD 905A are relevant UK specifications.

The ease of application and the properties of the deposits enable nickel to be used extensively in electroforming, Fig.4.4, for instance in making tools such as moulds and dies (which are themselves the source of other engineering products) or foil and mesh products.

4.4 *Miniature nickel bellows made by electroforming.*

Characteristics
1 Moderately hard, *i.e.* 150-450HV.
2 Softer and more ductile than chromium and so more resistant to fretting corrosion.
3 Resistant to corrosion in that 125µm ensures protection in most chemical environments.
4 Resistant to corrosion fatigue.
5 Resists scaling up to 600°C.
6 No technical limit to deposit thickness.
7 Deposits can be finished by turning.
8 Internal stress in deposits generally controllable.
9 Tends to gall when running against some metals even when lubricated.

Applications
1 Hydraulic equipment especially in marine service.
2 Processing of sensitive products such as food and textiles to avoid contamination, *e.g.* from rust staining.
3 Reclamation of worn parts.
4 Electroforming.

ELECTROLESS NICKEL

The deposits produced are essentially nickel alloys containing either 6-12% phosphorus or about 2-7% boron depending upon the reducing agent used. It is possible to produce ternary alloys, such as nickel-phosphorus-

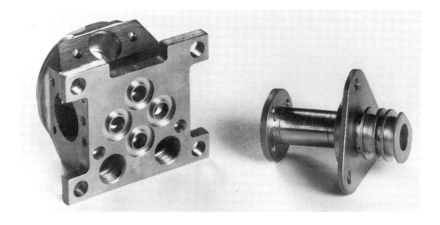

4.5 Hydraulic components electrolessly plated with nickel-phosphorus alloy (courtesy W Canning Materials Ltd).

copper and also to make composite coatings by suspending non-metallic particles in the deposition solution. DEF STAN 03-5/1 covers the requirements of nickel-phosphorus deposition, Fig.4.5.

Characteristics — nickel-phosphorus (Ni-P)
1. Uniform thickness.
2. Components can be plated to size without any finishing operation.
3. Hardness 400-500HV as deposited, can be raised to 1000HV by heat treatment at 400°C. (Properties are then generally comparable with hard chromium.)
4. Resistance to both corrosion and wear is superior to electrodeposited nickel.
5. Low frictional coefficient.
6. Retains room temperature strength up to 350°C and regains some strength on cooling from higher temperatures.
7. Difficult to plate high alloy steels, which may require an initial coating of electrodeposited nickel.
8. Materials more expensive than for electrodeposition.
9. Pre-treatment and solution control more critical than in electro-deposition.

Applications
1. Components for water pumps.
2. Hydraulic and pneumatic valves.
3. Plastics moulds.
4. Reclamation of worn parts.

Characteristics — nickel-boron (Ni-B)
1. Generally similar to Ni-P, but harder, *i.e.* 500-750HV.
2. Can be heat treated, if required, to increase hardness.
3. Less ductile and higher internal stress than Ni-P deposits.
4. Resists wear better than Ni-P deposits.
5. Operating costs greater than for Ni-P deposition.

Applications
As for Ni-P deposits, but Ni-B preferred when hot hardness important, *e.g.* in coating dies for glass moulding.

Copper
Copper electrodeposits, which are generally softer than nickel, can be applied easily from several well established electrolytes. The metal can also be deposited electrolessly and so is used extensively in the electronics industry. Copper can be electroformed to produce components such as spark erosion tools and wave guides for radar technology.

Characteristics
1 Soft and ductile. Hardness 60-150HV.
2 Good conductor of electricity.
3 Resistant to fretting corrosion.

Applications
1 Stop-off for selective case hardening of steels.
2 Printing rollers.
3 Surface lubricant in metal working.
4 Electroforming.
5 Reclamation of worn parts.

Copper-tin alloys
Alloying of copper with 10-15% tin produces compositions which are harder than either constituent metal.

Characteristics
1 Hardness 300-400HV.
2 Relatively ductile.
3 Resistant to corrosion and wear.
4 Resistant to shock loading.

Applications
1 Anti-sparking coating for underground pit props.
2 Undercoat for hard chromium.
3 Stop-off for selective nitriding of steels.

Iron
Electrodeposited iron has been used from the earliest days for reclamation of worn parts, *etc*, but its use appears to be considered seriously only when nickel is in short supply. Electroforming may be applied to make iron foil and also hollow roller bearing elements in iron-nickel alloys.

Characteristics
1 Inexpensive and strong.
2 Capable of heat treatment.
3 Not resistant to corrosion.
4 Not easy to deposit because of generally high temperature required and consequent oxidation of electrolyte constituents.

Applications
1 Soldering iron tips.
2 Anti-scuff coatings in IC engines.
3 Reclamation of worn parts.
4 Electroforming.

Cobalt
Unalloyed cobalt is used in computer technology because of its magnetic properties. However alloys with up to 20% of either molybdenum or tungsten have important engineering applications, notably in improving the wear resistance of metal forging and cold pressing tools, Fig.4.6. As these components are generally massive, and downtime must be minimal, the brush method can be used for plating in situ.

Characteristics
1 Hardness about 600HV.
2 Low frictional coefficient.

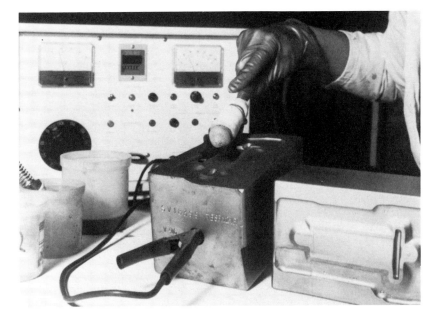

4.6 *Electrodeposition of cobalt-molybdenum alloy on to hot forging die by brush plating (courtesy Aston University).*

3 Resistant to oxidation.

4 Deposits as thin as 12μm are resistant to erosion and metal pick-up.

Applications
1 Hot forging dies.
2 Cold pressing tools.
3 Solid lubricants.

Lead
The importance of lead in engineering applications lies in the range of alloys which, when applied as overlay coatings, improve the performance of plain bearings. The alloys may contain 10% tin, 8% indium or 10% tin plus 2% copper.

Characteristics
1 Soft and ductile, hardness 8-15HV.
2 Low frictional coefficient, resistant to scuffing.
3 Resistant to corrosion by degraded lubricating oils.
4 Thickness about 25μm.

Applications
Overlay coatings for copper and aluminium-based plain bearings, to resist the severe conditions associated with starting and stopping of machine movements and to entrap small particles of dirt, so avoiding damage to bearing or shaft.

Tin
Like lead, electrodeposited tin is soft and its engineering applications are generally concerned with adhesive wear problems. BS 1872 and DEF STAN 03-8/1 cover deposition of tin.

Characteristics
1 Soft and ductile, hardness about 12HV.
2 Low melting point.
3 Low frictional coefficient, resistant to scuffing.
4 Solderable.
5 Resistant to corrosion.

Applications
1 Stop-off in nitriding of steel.

2 Corrosion protection in absence of mechanical wear.

3 Flash coatings to facilitate running-in of moving parts.

Characteristics — 65/35 tin-nickel alloy (covered by BS 3597)
1 High hardness, *i.e.* 700HV
2 Low frictional coefficient.
3 Resistant to corrosion.
4 Retains oil films.
5 Brittle, limited resistance to impact.
6 Upper limit about 360°C.
7 Solderable.

Applications
Automotive braking systems.

Tin-based diffusion alloys
In a fairly recent development, some tin-rich alloys are electrodeposited on to steel and then heat treated to form diffusion alloys containing iron-tin and other hard intermetallic compounds interspersed in a softer supporting matrix. Electrodeposited bronze diffusion coatings may also be applied to steel and cast iron.

Characteristics
1 Hardness 600-950HV.
2 Low frictional coefficient.
3 Resistant to corrosion.
4 Resistant to wear when lubricating conditions poor.
5 Minimal reduction in fatigue strength.

Applications
1 Pump plungers and bearings.
2 Textile thread guides.
3 Punches and dies for stamping and cutting stainless steel.

Silver
Electrodeposited silver has mechanical, electrical and chemical properties which are useful in many industries. Alloys with lead, palladium or copper extend the upper temperature limit. BS 2816, DEF STAN 03-9/1 and DTD 939 cover the requirements of silver deposition.

Characteristics
1 Hardness 60-120HV.
2 Thickness ranges from 0.001-1.0mm.
3 Resistant to fretting corrosion.
4 Good frictional properties.
5 Good conductor of electricity.

Applications
1 Special bearings, *e.g.* aircraft fuel systems.
2 Electrical contacts.
3 Combatting high temperature seizure.

Gold
Because of its unique combination of properties, electrodeposited gold finds extensive use in electronics, control and communications equipment. The pure metal is relatively soft, *i.e.* about 70HV and is free of surface oxide films so that galling may occur between mating surfaces, *e.g.* in sliding contacts. However, the incorporation of a small proportion of base metals, such as

nickel and cobalt as alloying constituents increases hardness to about 450HV. This factor, combined with the co-deposition of an organic polymer generated in the plating solution, leads to an improvement in wear resistance and so extends the range of applications of alloy gold coatings for electrical contact purposes.

Characteristics
1 Hardness of pure gold, *i.e.* about 70HV can be increased by alloying with base metals, such as nickel and cobalt.
2 Low electrical contact resistance.
3 Withstands corrosion and oxidation over a wide temperature range.
4 Good solderability, which is retained even after long storage.

Applications
1 Edge connectors in printed circuit boards.
2 Transistor header components.
3 Connectors in electrical and electronic equipment.

Platinum group metals
Electrodeposited coatings of palladium, platinum, rhodium and ruthenium are of practical engineering interest because of their hardness and resistance to corrosion. Thus thin coatings are applied to protect components against wear and corrosion in electrical and electronics equipment. The main characteristics and applications are summarised briefly below.

Palladium
Palladium may be deposited at relatively low cost and its hardness of about 300HV commends it for use in heavy duty electrical contacts. Palladium coatings are also used to protect refractory metals, such as molybdenum, against oxidation.

Platinum
Platinum is especially resistant to corrosion. Very thin coatings applied to titanium produce electrodes which are widely used for cathodic protection purposes and in industrial electrochemical processes.

Rhodium
Rhodium is easily deposited and has a hardness of about 800HV. It is used in manufacture of rotary contacts operating in conditions of extreme wear and is also employed to protect silver against tarnishing and against wear in contacts carrying high currents.

Ruthenium
Ruthenium is becoming of increasing interest and may replace rhodium in certain applications.

Metal-ceramic composites
The ability to co-deposit hard particles extends considerably the range of wear resistant coatings applicable by electrodeposition. Commercially available processes are based on cobalt and nickel matrices containing about 30% by volume of particulate materials (around 3μm size), such as chromium carbide and silicon carbide. Other matrix metals may be applied and softer materials co-deposited, for example PTFE for lubrication purposes. Electroless deposition methods may be modified to produce composite coatings with the advantage of uniform thickness. DTD 943 covers deposition of cobalt-chromium carbide composite coatings, Fig.4.7.

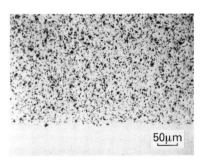

4.7 *Typical microstructure of a cobalt-chromium carbide electrodeposited composite coating (courtesy BAJ Vickers Ltd).*

Characteristics — cobalt-chromium carbide
1 Hardness 350-500HV.
2 Resistant to oxidation and fretting up to 800°C.

Applications
1. Aircraft jet engine components.
2. Textile machinery components.

Characteristics — nickel-silicon carbide
1. Hardness 350-500HV.
2. Wear resistance not generally as good as cobalt-chromium carbide.

Applications
1. Rotary IC engines.
2. Seals for operation in fresh and seawater.
3. Moulds for forming glass.

Hard anodising

Aluminium alloys can be easily electroplated, for example with chromium, to improve wear resistance, but hard anodising constitutes a useful alternative method. Although applicable to most alloys, those containing more than about 3% copper or 7% silicon generally require modified treatment. BS 5599 and DEF STAN 03-26/1 are the relevant UK specifications.

Characteristics
1. Hardness 400-500HV.
2. Wear resistance comparable with hard chromium.
3. Thickness 25-75μm, sometimes up to 175μm.
4. Porosity may be sealed with lubricants.
5. Brittle, not resistant to shock loading.
6. Can be sealed to improve corrosion resistance, but wear resistance is reduced.

Applications
1. Transport industries especially aerospace.
2. Hydraulic valves.
3. High speed machinery.
4. Injection moulds as an economic alternative to steel.
5. Self-lubricating surfaces.

Selection

Wear is a complex phenomenon and in practice more than one form is generally operating (see Chapter 2). For example, the surfaces of components may be subjected to relative movement in a corrosive fluid containing an abrasive powder. In this case, a designer would choose a hard coating material, such as chromium, to resist both abrasion arising from scoring or erosion and corrosive attack. A low coefficient of friction is desirable to combat adhesive wear associated with scuffing or galling, and chromium or electroless nickel would be natural choices. Softer coatings, however, such as electrodeposited nickel or copper, are preferred to combat fretting corrosion.

The development of wear resistant coatings in which hard particles are co-deposited with a matrix metal, generally Ni or Co, adds to the design potential of this surfacing process. For wear resistance the hard particles are often carbides of chromium or silicon, but softer materials, such as molybdenum disulphide for lubrication, can also be used. Electroless deposition methods may also be modified to produce composite coatings of uniform thickness. In selecting a coating the designer may also have to consider factors such as ease of machining and resistance to shock loading.

Table 4.3 indicates an empirical order of suitability of the more usual electrodeposited engineering coatings. The ratings quoted, which are generalised and subjective, serve only as an approximate guide and it is important to consult a specialised industrial electroplater before reaching a decision on a suitable coating.

Table 4.3 Guide to selection of an electrodeposited engineering coating

Coating	Relative cost* (A = least costly)	Ease of deposition	Ease of machining	Abrasion resistance at normal temperatures	Protection against corrosion	Resistance to shock loading
Chromium	D	3	1	6	1	2
Electrodeposited nickel	A	6	6	1	5	6
Electroless nickel −9% phosphorus	F	5	5	4 (heat treated)	6	4
65/35 tin-nickel	B	4	3	3	4	1
Cobalt-chromium carbide composite	E	1	4	2	3	5
Hard anodised aluminium	C	2	2	5	2	3

Note: 1 = lowest or worst, 6 = highest or best
  * Costs depend on many factors, such as the shape of the work and the volume of production, especially for electroless nickel coatings.

Design

Electrodeposition must be regarded as an integral part of the production process and the designer should consult the electroplater at an early stage to achieve the best results.

The characteristics of an electrodeposited coating are primarily determined by the following:

1 The electrolyte system in use;
2 The lack of local uniformity of current density on the workpiece;
3 The dependence of deposition rate on current density, deposition time and deposition efficiency.

Other general features include the inability of electroplating to fill holes, level rough material and cover defects satisfactorily. While it may be possible to deposit some plating over such defects, the deposit will be poorly bonded to the substrate. Furthermore, cracks or surface depressions tend to trap air or gas bubbles which prevent effective solution control and result in poor plating.

The electrolyte system will generally not be under the control of the designer, but he/she can do much to alleviate the problems arising from the other factors, and can be helped by good electroplating practice. These two aspects are considered separately.

Component design

Figure 4.8 shows examples of component design features that avoid or minimise many of the undesirable effects generated by electrodeposition. The underlying principles are to avoid sharp edges, sharp changes of section, close spacing of members, protruberances, and traps for solution or gas bubbles.

When plated parts have to be fitted together, allowance must be made for coating thickness and variations in this thickness. Screw threads, for example, present a problem, and specification of an electroless process, if possible, improves the uniformity of the deposit.

As the part being plated must be connected electrically into the circuit, the component must be suitably held. Sometimes this is best achieved by designing lugs away from the working surface which can be removed easily after plating. The lugs may also facilitate holding for surface preparation and polishing, Fig.4.9.

As insoluble particles in the solution may settle on the uppermost surfaces of the component (covering roughness and possibly lower corrosion resistance) the designer should arrange either for non-significant surfaces, or for areas of the significant surface which are of less importance, to be uppermost when the article is in the plating solution.

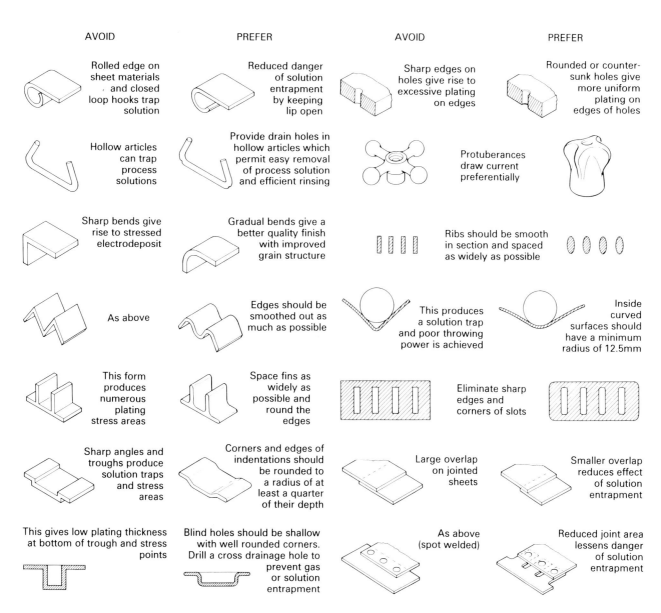

4.8 *Design guides and requirements for components to be electroplated.*

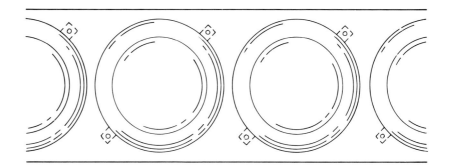

4.9 *Placement of temporary holding lugs.*

Although barrel plating avoids many of the constraints of vat plating, it introduces some special requirements. For example, design considerations should take account of the non-uniform distribution of finish in a barrel, caused by masking effects between adjacent parts. Parts must not be prone to interlocking. Flat parts such as washers can be dimpled to prevent sticking. Hollow parts may not immerse easily so they should be provided with drainage holes.

Process design for vat plating

Deposit thickness is dependent on current density and so will be greater at edges, corners and projections but less at the centres of large, flat areas and in surface recesses. Thus, even a flat sheet will not have a uniform current distribution and hence will not receive a uniform deposit. Figure 4.10 shows the current distribution, and the resulting coating thickness, which results from a simple, uncompensated, arrangement of anodes.

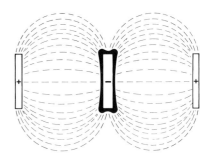

4.10 *Distribution of current and electrodeposit on a cathode of rectangular cross section.*

Good plating practice can overcome, the problem of deposit uniformity to a large extent by making the current distribution as uniform as possible, *e.g.* by:
1. Correct positioning and size of anodes;
2. Using anodes which approximate to the shape of the surface (or part) to be treated;
3. Positioning 'burners' or 'robbers' (electrically connected to the cathode) near to high current density areas and so reducing the amount deposited on those areas;
4. Using non-conductive shields to reduce current flow at sharp edges.

The diagrams in Fig.4.11 illustrate these principles.

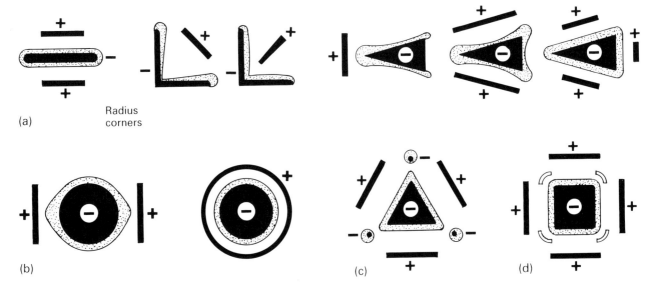

4.11 *Plating practices to aid deposit uniformity:* a) *Positioning and size of anodes;* b) *Conformal anodes;* c) *Robbers;* d) *Non-conductive shields.*

Substrate materials

Electroplated coatings can be applied to all the metals, and most of the plastics, used in engineering, but care is needed with certain materials. The constraints on plastics substrates are concerned with process temperatures and compatibilities, but some steels are susceptible to the hydrogen that is released as a competing process to the electrodeposition of metals. Hydrogen ions in the electrolyte are plated out as atomic hydrogen which may rapidly diffuse into steel workpieces. If the steel has a tensile strength exceeding 1000 N/mm^2, or if it has been heavily cold marked before plating, it may consequently suffer from hydrogen embrittlement. Heat treatment at 200-250°C after plating should release the hydrogen. It is preferable, however, to stress relieve components before electroplating if possible.

Surface preparation

Substrate surfaces should be sound and without significant pores or defects such as cracks. The surface should be clean, smooth, and free from burrs and machining or polishing debris. Traps for debris are best avoided by current design. A polished or fine mechanical finish is most suitable.

Precleaning involves degreasing by solvent and/or an alkaline electrolytic bath, and so the components should be compatible with these processes, but the major responsibility for precleaning lies with the electroplater.

To avoid unnecessary plating, those areas not requiring treatment may be suitably stopped-off by the use of lacquer, tape, a well fitting mask, or by screen printing, again, the work should be compatible with the chosen method.

As in all aspects of coating, consultation between the customer and electroplater is essential if problems are to be avoided.

Finishing

Reshaping of plated parts must be avoided or the coating will be damaged, but some finishing, such as polishing, can be done if necessary. Ideally however, the finish on the coating closely resembles that on the original workpiece. For close tolerance work, however, machining is necessary.

Electrodeposits are usually in a state of stress which can reduce fatigue strength if the deposit is in tension. In stressed parts, such as aircraft components, coating tensile stresses may be countered by introducing compressive stresses in the substrate by shot peening before plating. Alternatively stresses can be reduced by heat treatment, provided that the mechanical properties of the coating and substrate are not unduly influenced. Coatings on plastics substrates cannot be stress relieved by the above means, and so the electrodeposition process must be chosen to minimise residual stresses.

Specifications, inspection and quality assurance

The designer should specify significant surfaces on components, masked areas, the material to be deposited, thickness distribution and tolerances, coating hardness and other mechanical properties, surface finish to be achieved, appearance of the coating, and other specific requirements. The electroplater will advise on product shape and finish, and on requirements for cleaning, jigging, plating, *etc.*

A procedure for sampling and inspection to meet the above should be agreed with the electroplater, and a procedure for handling items that fail inspection should also be agreed. Many of the standards quoted in this chapter give detailed guidance on specification and inspection, but it must be emphasised that non-destructive testing of thin coatings (of whatever nature) is limited, and so quality assurance must play a major role in production of any component that is to be used as a critical part.

Physical and chemical vapour deposition techniques

CHAPTER 5

The processes

This chapter covers methods for producing overlay inorganic coatings which are formed on the surface of a substrate by condensation or reaction from the vapour phase. It does not cover thermal or thermochemical diffusion treatments, but ion implantation (although strictly not a coating process) is included.

Physical vapour deposition (PVD) coating techniques are almost wholly confined to making relatively thin films (ranging from 10^{-7} to 10^{-4}m) whereas chemical vapour deposition (CVD) is used both for thin films and for coatings in excess of 1mm. Such coatings have an important industrial role because of the anti-wear and anti-corrosion properties that they confer on engineering components, and in addition to these uses, there are large markets in the field of electronic and optical devices, product decoration, and in architecture. As a result, development is taking place in all these directions and, as each application emphasises its own special set of requirements, hybrid techniques are continually emerging; in fact, the present extensive overlap in PVD and CVD methods means that hard and fast distinctions are difficult to make. The commercial success of many of the variants is open to doubt because their characteristics are not yet fully known and they have not yet found their niche in the market. The processes and their characteristics are summarised in Table 5.1.

Table 5.1 Comparison of process characteristics

	Processing temperature, °C	Throwing power	Coating materials	Coating applications and special features
Vacuum evaporation	Room temperature (RT)→ 700 usually < 200	Line of sight	Chiefly metal, especially Al (a few simple alloys/ a few simple compounds)	Electronic, optical, decorative, masking simple
Ion implantation	200-400 best < 250 for N	Line of sight	Usually N (B, C)	Wear resistance for tools, dies, etc. Effect much deeper than original implantation depth. Precise area treatment, excellent process control
Ion plating, ARE*	RT→ 0.7Tm† of coating. Best at elevated temperatures	Moderate to good	Ion plating: Al, other metals (few alloys) ARE: TiN and other compounds	Electronic, optical, decorative. Corrosion and wear resistance. Dry lubricants. Thicker engineering coatings
Sputtering	RT→ 0.7Tm of metal coatings. Best > 200 for non-metals	Line of sight	Metals, alloys, glasses, oxides. TiN and other compounds (see below)	Electronic, optical, wear resistance. Architectural (decorative). Generally thin coatings. Excellent process control
CVD	300-2000 usually 600-1200	Very good	Metals, especially refractory TiN and other compounds (see below) pryolytic C, pyrolytic BN	Thin, wear resistant films on metal and carbide dies, tools, etc. Free standing bodies of refractory metals and of pyrolytic C or BN

*Activated Reactive Evaporation
(includes sputter ion plating and arc evaporation)
† Tm = absolute melting temperature
Compounds: oxides, nitrides, carbides, silicides, borides of: Al, B, Cr, Hf, Mo, Nb, Ni, Re, Si, Ta, Ti, V, W, Zr

All vapour processes involve treatment in a chamber — either in a vacuum chamber or in one which can withstand the high temperature and corrosive gases which are used in CVD. This limits somewhat the size of object to be coated but this limitation is largely imposed by the capital expenditure involved rather than by any fundamental characteristics of the process. Providing that the substrate can be manipulated to face the coating source, the size and shape of objects to be coated is again only limited by capital expenditure on plant. Small objects are held in large numbers of baskets and barrel-coated as in electroplating; at the other end of the scale, large objects such as aircraft undercarriage units are coated with aluminium in chambers

5.1 a) *Continuous sputtering line for solar coatings on architectural glass;* b) *The Motley No 2 Building, Texas, USA (courtesy Airco-Temescal, BOC Technologies).*

5.2 *Plant for vacuum evaporation of aluminium on to plastic foil (courtesy Leybold-Heraeus).*

of 2m diameter by 3m long, and window glass sheets up to 6m by 8m are solar coated in plant up to 60m long, Fig.5.1.

Most processes are operated on a batch basis, Fig.5.2, but vacuum evaporation on to paper or plastic sheet may be run with a continuous air to air feed through differentially-pumped inlet and outlet ports.

The PVD processes described in this chapter are those which illustrate the major factors governing PVD characteristics and the nature of the coatings that they produce; they are also the most widely used. PVD coating vapours are generated either by evaporation from a (usually) molten source, or by ejection of atoms from a solid source which is undergoing bombardment by an ionised gas (*i.e.* sputtering). The vapour may then be left as a stream of neutral atoms in a vacuum (vacuum evaporation) or it may be ionised to a greater or lesser extent. A partially (say 0.1%) ionised stream is usually mixed with an ionised gas and deposits on an earthed or biased substrate (ion plating and sputter coating), but a highly ionised stream which forms a plasma is attracted to a biased substrate (arc plasma evaporation). Alternatively, a 100% ionised beam may be focused and accelerated to sufficiently high energies to penetrate into the substrate (ion implantation). This chapter concentrates only on those processes which are in commercial use, Fig.5.3. The section on CVD coatings is subdivided on the basis of the chemical reactions involved.

SURFACE PREPARATION

No vapour deposition method gives an adhesion acceptable for engineering purposes unless the substrate is truly clean. The standard of cleanness

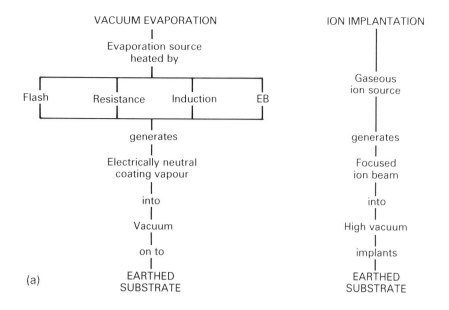

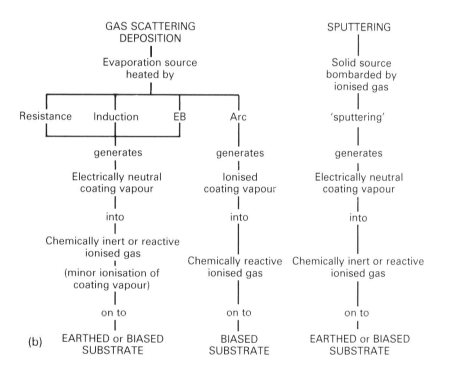

5.3 *Methods of physical vapour deposition:* a) *Vacuum deposition processes;* b) *Gas scattering processes.*

required is far beyond that needed for weld surfacing or for spraying, and it involves removal of contaminant layers only a few tens of molecules thick. Such preparation cannot be done in air; the most appropriate technique is ion bombardment and it is assumed that many of the PVD processes to be described will have been preceded by such a cleaning cycle. Ion bombardment cleaning is performed by a low pressure gas discharge; the work chamber contains argon at a pressure of some 10^{-2} torr and the workpiece is made negative at 2-5kV with respect to earth. Positive argon ions generated in the discharge are accelerated to the workpiece at high energies and eject surface atoms when they arrive. The substrate is thus cleaned by an erosion process which 'sputters' contaminant and substrate atoms into the chamber.

CVD processes, on the other hand, normally require only that the substrate be properly degreased and is free from obvious oxide films; a cleaning cycle involving a reducing gas at elevated temperatures is often all that is required before coating begins.

Vacuum evaporation

Vacuum evaporation is probably the most widely used of PVD techniques and it accounts for the major proportion of both the equipment in use and the area coated. It is also the oldest of the PVD methods and is, in principle, the simplest, but it is rather limited in its applications. Some of these limitations can be reduced by using coating material sources that ionise their output, but some methods are much more complex and cannot yet be said to represent a significant industrial process.

Vacuum evaporation is usually conducted in a hard vacuum of 0.1-10mPa, at which pressures the mean free path of a gas atom (*i.e.* the average distance travelled before it suffers collision with another atom) is about 1-100m. As this is very much greater than the chamber dimensions, an atom evaporating from a source will travel in a straight line and so the process is essentially one of line of sight; coating around corners or into re-entrants is not possible without physical movement of the substrate, Fig.5.4.

5.4 *Rotating jig plant for coating car headlamp reflectors by vacuum evaporation (courtesy Leybold-Heraeus).*

Coating material vapour is usually produced by thermal evaporation. The vapour consists largely of single atoms or clusters of atoms although in some variations of the process there may be an appreciable degree of ionisation of the vapour stream. Except where ionisation is present, the energy of the vapour particles is relatively low and is merely that necessary for evaporation (around 0.1-1eV).

The substrate to be coated is usually unbiased (*i.e.* earthed) and may be heated or cooled. Deposition rates in vacuum evaporation are amongst the highest in all PVD processes and can range up to 75 μm/min, but a more usual figure is around 2 μm/min. Coatings can be deposited to preserve the finish of the underlying surface, but adhesion of thermally evaporated coatings is wholly dependent on the cleanness of the substrate immediately before deposition starts. For many less demanding industrial applications, the ion bombardment cleaning cycle is omitted and adhesion consequently suffers.

EVAPORATION SOURCES

Vacuum evaporation at sensible rates implies a vapour pressure of the source material of about 1Pa which, for the majority of materials, requires a temperature well above the melting point. Thus some form of crucible must be used that can contain the melt without reaction and this can be a major problem. Heating the source is commonly accomplished by direct resistance heating of, *e.g.* tungsten or molybdenum boats, Fig.5.5, by induction heating, or by electron beam heating of the material in a water cooled copper crucible, Fig.5.6. A major exception is aluminium which is evaporated from an electrically-conducting refractory boat of boron nitride/titanium dioxide mixture because the molten metal reacts with tungsten, *etc.*

Within limits, alloy deposition is possible if account is taken of the different vapour pressures exerted by the alloy components. Although the problem

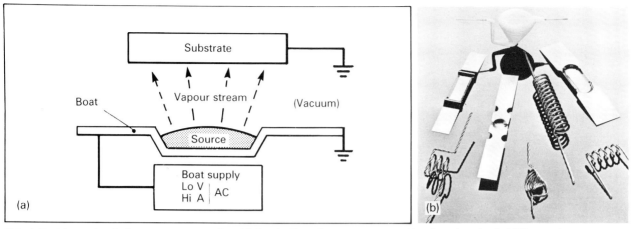

5.5 a) *Resistance heated vacuum evaporation;* b) *A selection of vacuum evaporation (courtesy Leybold-Heraeus).*

can sometimes be avoided by dropping a fine powder of the alloy on to a super-heated boat to cause total vaporisation without fractionation (flash evaporation), limitations are imposed by reactions between the powders and the boat.

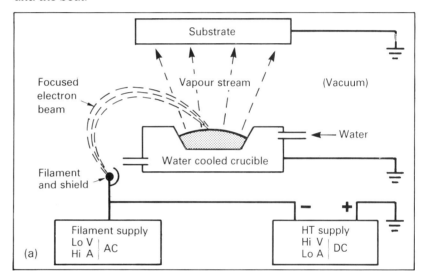

5.6 a) *Electron beam heated vacuum evaporation (270° deflection);* b) *Electron beam heated, water cooled copper crucibles for vacuum evaporation, 90° deflection, 25kW (courtesy Leybold-Heraeus).*

Electron beam heating is a form of skull melting and so avoids the problem of reaction between the molten source and the crucible. It can also be used for making alloy coatings by oscillating the beam between crucibles containing the individual constituents and correspondingly adjusting beam power to produce the correct vapour pressure above each. However, the beam generator must be shielded from the metal vapour, and the systems are complex, Fig.5.6.

MATERIALS AND APPLICATIONS
The major applications of vacuum evaporation are confined to coatings of single metals such as aluminium, (which constitutes 90% of all materials

evaporated), chromium, silver, *etc*; few compounds can be evaporated without change of composition.

Engineering uses of vacuum coatings are limited — general aerospace applications use Al and Ni-Cr (for corrosion protection), and Al and Ag as solid lubricants. Hard coatings of Cr or Al_2O_3 are sometimes deposited on steel or tungsten carbide tools, but this application has now largely been taken over by sputter coating. Probably the largest area coated is for decorative purposes, *e.g.* Al and Cr for automotive trim, Al for packaging and costume jewellery, and fluorides for iridescent effects. Evaporated coatings for electronic applications include Al and Ta on to plastics film for capacitors, Al-Si and Ni-Cr by flash evaporation for thin film resistors, and Si, GaAs and CdS for semiconductors and solar cells respectively. Optical coatings include Al and Ag for mirrors and reflectors and also for solar coatings on windows; in addition, evaporated MgF_2 is well known as an anti-reflection coating on lenses.

Gas scattering deposition

Vacuum evaporation, being a line of sight process, normally has poor 'throwing power'. However, the presence of gas molecules in the coating chamber reduces the mean free path of the evaporant atoms, and, if there are sufficient collisions, the coating atoms will no longer impinge on the substrate in a straight line from the source. At 1Pa, the mean free path is about 5mm; thus, in a system with a typical source to substrate distance of some 200mm, evaporant atoms will suffer over 50 collisions en route and will arrive at the substrate from almost any direction; their arrival route will now be partially controlled by a gas diffusion process. The pressure range for this regime is from 0.1-5Pa at which there is a substantial gain in throwing power without appreciable reduction in evaporation rates; furthermore, the introduction of a gas into the coating chamber offers a number of new variables which allow an improvement in coating quality and an extension of the number of materials which can be deposited as coatings. The major variables are: electrically neutral or ionised gas, and chemically inert or reactive gas — the combinations are illustrated in Fig.5.7.

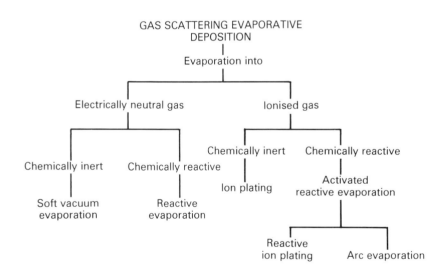

5.7 *Gas scattering evaporative deposition.*

ION PLATING

General principles

Evaporation into an electrically neutral gas has been virtually superseded for industrial purposes by use of an ionised gas as the scattering medium, *i.e.* ion plating. A dynamic pressure of some 1Pa of, *e.g.* argon, is maintained by a continuous gas bleed and throttled pumping (a dynamic pressure is preferred to prevent the build-up of outgassed impurities in the chamber), and the substrate is negatively biased at 1-2kV with respect to the grounded evaporation source, Fig.5.8.

Evaporant atoms gain energy by collision with the ionised argon but, contrary to earlier belief which originated the name ion plating, the degree of ionisation of the vapour stream is probably no more than 0.1%; the term ion

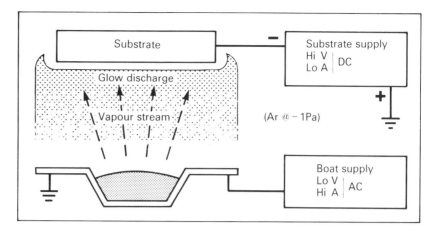

5.8 *Resistance heated ion plating.*

plating is therefore largely a misnomer but is widely accepted. The combination of high energy (approximately 10eV) evaporant atoms plus continuing substrate bombardment by gas ions results in excellent film adhesion and, by correct choice of deposition parameters, in a dense coating. At the same time, the presence of the gas ensures a good throwing power, Fig.5.9.

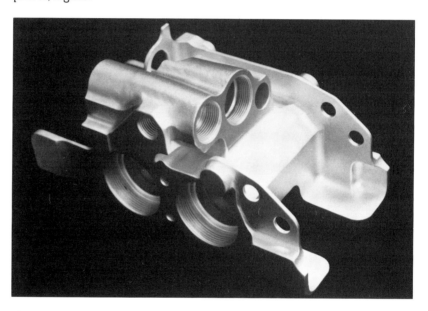

5.9 *Complex aircraft casting coated with aluminium by ion plating (courtesy General Engineering).*

The evaporant charge is melted as in vacuum evaporation, and gas ionisation may be advantageously increased by, *e.g.* a heated filament electron emitter. There are several variants of the technique and all claim high film adhesion and density, but all require the substrate to be held at a high DC potential or to be connected to an RF supply.

Materials and applications
Deposition rates are somewhat lower than in vacuum evaporation (some coating material is scattered away from the substrate) but they are nevertheless amongst the highest of the PVD techniques. Dense deposits are best achieved by deposition at elevated temperatures up to 0.7Tm and the grain structure of the deposit (columnar, equiaxed, *etc*) is controlled by a suitable choice of deposition rate, gas pressure, substrate bias and substrate temperature.

Ion plating excited considerable interest over the past twenty years but has now settled down to a number of important industrial applications. As with vacuum evaporation, the method is limited to those metals that can be evaporated, and the same limitation with regard to crucibles, *etc*, apply. Furthermore, because the substrate holder — which may be required to oscillate and rotate the parts to be coated — is maintained at a high potential, the complexity of the equipment is greater than that for simple vacuum evaporation, nevertheless the improvement in coating quality is such as to warrant that complexity.

It is not practical to electroplate aluminium for corrosion protection, and thermally sprayed aluminium may not be suitable for some components, but ion plated films of aluminium are routinely applied to complex aircraft parts, Fig.5.9. Ion plating rivals or has even displaced electroplating for some materials such as cadmium where there is a danger of hydrogen embrittlement of high strength steel substrates, and where the poisonous nature of the solutions has caused them to be banned in some countries; chromium is increasingly being deposited on to plastics for car trim for the same reason. Solid lubricant films of, *e.g.* silver, for aerospace applications are ion plated on to bearings, *etc*, that are to be exposed to heat, radiation and hard vacuum.

In general, any material that can be vacuum evaporated can be ion plated with superior adhesion and density.

ACTIVATED REACTIVE EVAPORATION TECHNIQUES (ARE)

Reactive ion plating

Vacuum evaporation and ion plating are both restricted to coating materials that can be evaporated without change of composition. There is a considerable demand, however, for wear resistant coatings of oxides, nitrides, carbides, *etc*, and it was for this reason that glow discharge ARE (also known as reactive ion plating) was developed. The principle is similar to that of ion plating except a reactive, instead of an inert, gas is used to generate the discharge. The metallic constituent of the compound is evaporated into the reactive gas whose reactivity is considerably enhanced by the glow discharge, and reaction takes place either in the gas phase, or more commonly, on the substrate surface. Thus, a thin, hard, wear resistant coating of titanium nitride (TiN), for example, can be produced by evaporating titanium into a low pressure glow discharge of a mixture of argon and nitrogen; one variant of this technique is shown in Fig.5.10.

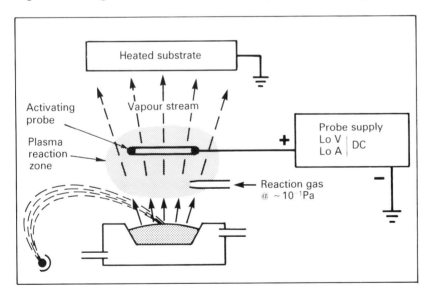

5.10 *Reactive ion plating.*

The diagram shows what is basically an electron beam heated vacuum evaporator into which the reactive gas is introduced close to an auxiliary ionising electrode. The resultant plasma activates the reaction by ionisation of the reactive gas and by creation of metastable species. Further complexity is introduced by the necessity to control the auxiliary electrode and the balance of the reactive and inert gases, but ARE is one of the two evaporative methods used for making coatings of many compounds.

Arc evaporation ion plating

The reactive methods described earlier use some form of indirect heating to melt the coating material which is evaporated with little ionisation into a generalised glow discharge. Arc evaporation, on the other hand, uses a solid metal target which is locally and momentarily struck between the source (cathode) and either the chamber (extended anode) or a localised anode, although unlike atmospheric arcs, no luminous discharge path is present. The high power density at the root of the arc ($>10^9$ W/m^2) causes an intense evaporation of metal vapour which (in contrast to ion plating and other forms of ARE) is extensively ionised and is attracted to the negatively biased substrate, Fig.5.11.

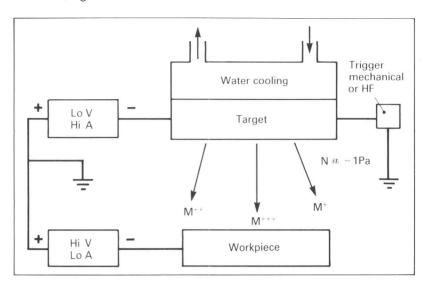

5.11 *Arc evaporative ion plating.*

The substrate undergoes a vigorous bombardment by metal ions which clean and heat the surface without causing a build-up of coating material. A reactive gas is then admitted and the parameters are adjusted to give ion plating conditions.

The use of an arc to cause evaporation from an effectively solid target overcomes many of the limitations of fully molten sources. Thus, high melting and reactive metals are melted without crucible problems, and multiple sources can be arranged vertically, horizontally downwards, or at any angle to improve the speed and uniformity of coating. The combination of elevated substrate temperature and highly energetic coating ions ensures excellent adhesion and dense coatings, but droplet ejection from the target may cause a higher degree of surface roughness on the coating than is found in other reactive techniques.

Sputter ion plating (SIP)

Another commercially important technique for overcoming the limitations of evaporative vapour production (and one which does not involve any melting whatsoever of the coating source) is ion bombardment sputtering which is more fully described later in this chapter. Sputtering does not require the coating source to be melted and it therefore adds considerably to the range of materials that can be deposited. The SIP process can be run with both inert and reactive gases, but it is much slower than the evaporative techniques. On the other hand, the characteristics of SIP include high purity and control of the constituents and components in the system. Furthermore, it is amenable not only to large chamber operation, but also to treating mixed batches of components; it is these factors, particularly, which give the technique a certain commercial advantage over its competitors.

Materials and applications

Oxides, carbides and nitrides are the principal compounds described by SIP and the various ARE techniques by matching the generation rate of the metallic constituent with the partial pressure of oxygen, ethane or nitrogen respectively. Molybdenum disulphide films are also produced, and other compounds such as borides or silicides have been applied as coatings, but there is, as yet, no major industrial use for these latter.

The compound that dominates this field is titanium nitride. Films of TiN, some 5μm in thickness, are now applied to bearings and to a wide range of tools such as simple twist bits, complex gear hobs, punches, dies, taps, milling cutters, forming tools, *etc*, Fig.5.12. Tool life is improved by three to ten times by the coating and cutting speed can often be increased; moreover, the quality of the work is also improved.

5.12 *A selection of tools coated with TiN by ARE (courtesy Tecvac Ltd).*

The wear resistant films are golden coloured and are also used to provide decorative scratch resistance to, for example, stainless steel or gold watch cases. Substrate temperatures are generally held above 400°C for the more demanding tribological applications, but films for decorative purposes may be deposited at much lower temperatures.

Although TiN is currently by far the most widely used of the wear resistant thin film compounds, others such as titanium carbide, titanium carbonitride, titanium-aluminium nitride, and carbides or nitrides of, for example, hafnium, chromium or zirconium, are used on a much smaller scale. They may find specialised applications at, for example, temperatures above 500°C where TiN is unsuitable. Research into thin tribological coatings is presently very active, and undoubtedly new coatings, and multilayer coating systems, will emerge in the next few years.

In the optical/electronic fields, ARE and its associated techniques have allowed manufacture of scratch resistance anti-flare coatings of compounds that could be evaporated, and there is also an important market for thin (<0.1μm), transparent, and electrically conducting films of indium-tin oxide deposited as de-icing layers on to, for example, aircraft windows.

There is presently much emphasis on thin tribological films, but metal and — to a much lesser extent — alloy coatings are also made by these ionisation-assisted techniques. Sputter ion plating produces dense, adherent and pure coatings of the refractory metals, and aluminium and cadmium coatings for corrosion protection have been reported. Aluminium conductor tracks are also ion plated on some electronic devices, as are copper tracks. However, the major area of operation of all these techniques is to deposit thin coatings of materials — metals, alloys, compounds — that cannot be deposited by electroplating.

The various activated reaction techniques are in competition and there is much overlap between them, particularly for production of titanium nitride coatings, but it is too soon to tell which, if any, will eventually command the field. The different characteristics of the techniques have, in the past, led to different characteristics in coatings of nominally the same material, but these differences seem to be decreasing as the processes become better understood and controlled. The eventual dominance of one method may be influenced more by its commercial than by its technical characteristics.

Sputter coating
GENERAL PRINCIPLES
In the ion bombardment cleaning cycle that precedes evaporation coating, material is ejected from the face of the negatively biased substrate by argon-ion bombardment. The ejected material is in the form of high energy atoms which condense on the surrounding surfaces (the darkening that occurs at the ends of fluorescent light tubes as they age is material that has been sputtered from the electrodes on to the glass). Because the high energy of the bombarding ions is sufficient to overcome the binding energy of any substrate lattice, and because the source generates its vapour not by heating and evaporation but by ion bombardment, the technique can be used for depositing a wide range of coating materials (the source is called the target), Fig.5.13. It is this feature, above all others, that makes sputter coating the most versatile and probably the most widely used process for engineering, optical and electronic thin films.

The target is a solid and can be of an alloy or compound (or even a pseudo-alloy consisting of a mixture of the powdered constituents), the gas ionising potential is applied to it as DC or RF. The gas itself may be inert or reactive and the substrate may be earthed or biased; the combinations are shown in Fig.5.14. The temperature reached by the substrate is governed by many factors, but unless higher values are required, it is generally below 200°C.

There are snags of course. The necessary gas pressures (around 0.5Pa) reduce the mean free path to 30mm or so, therefore the substrate must be approximately at this distance if the high energy of the coating species (with all the advantages that this confers) is not to be lost. This in turn means that throwing power is compromised and that the target must be similar in size and shape to the substrate. There are other snags, but the most unfortunate

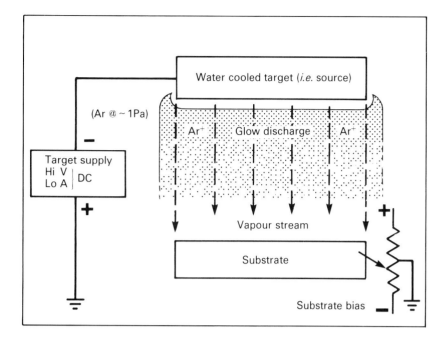

5.13 *DC diode sputtering.*

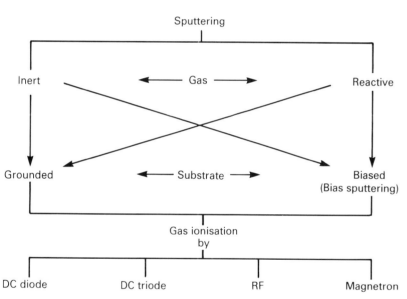

5.14 *Variants of sputter coating.*

is the low rate of deposition; the simple diode system shown in Fig.5.13 cannot exceed coating rates of 0.1 μm/min at best and with some coating materials it operates at one or two orders of magnitude lower. More complex systems involving auxiliary ionising sources may achieve 0.4 μm/min, but a slow deposition rate is not only unfortunate from a production aspect — it can also lead to a high gas content in the deposit. A film deposition rate of 0.1 μm/min is equivalent to about 3 monolayers per second and this is about the same figure as the arrival rate of molecular oxygen at a partial pressure of 1mPa. Good vacuum practice is therefore needed to maintain oxygen-free films; furthermore, hot deposition of subsequent heat treatment may be required to desorb occluded discharge gas. An important development however, is the magnetron source which can give very high rates of deposition. This is decribed later.

Despite its problems, the sputtering technique can deposit excellent (albeit thin) coatings of materials that cannot be deposited by any other method.

SUBSTRATE BIASING
The usual sputtering practice is to apply an ionising potential to the target only, leaving the substrate earthed, but there is a variant of the technique known as bias sputtering in which the substrate is also biased, but at a lower potential than the target. The effect of substrate biasing is to divert some of the ion bombardment to the substrate and so to gain the advantages that

apply to ion plating; possibly this variant should be regarded as a form of ion plating in which sputtering (rather than evaporation) is used to generate the coating material vapour. Substrate bias may be positive or negative, according to the ionised gas used and the composition of the film to be deposited.

GAS IONISATION

Sputter coating has been in use since the 1920s but in its early form, as illustrated in Fig.5.13, it had two major disadvantages, namely the necessity to confine its operation to electrically conducting targets (insulators became charged during bombardment and repelled gas ions) and a low coating rate (caused by a comparatively inefficient argon ionisation rate).

The problems of insulating targets becoming charged under DC bombardment are overcome by the use of RF excitation. If an insulated or non-conducting target is capacitively coupled to an RF generator the space charge that would otherwise build up is effectively abolished and the surface adopts negative charge. Frequencies above 10MHz are used and the RF bias is typically 2.5kV peak to peak. The target can therefore be of an insulating material and oxides, for example, can be directly sputtered on to a substrate although the erosion rate of an oxide is generally only about 10% of that of the corresponding metal. The ability to produce coatings directly from non-metallic targets is a valuable property of RF sputtering.

Early attempts to increase the ionisation efficiency of the simple DC diode system were successful to some degree and sputtering rates were increased by a factor of three or so, but the introduction of magnetically assisted (magnetron) sputtering led to deposition rates which are comparable with those from vacuum evaporation. Magnetron sputtering has undoubtedly been responsible for the widespread adoption of sputter coating by many branches of modern technology.

Magnets are arranged behind the target such that the magnetic flux lines lie parallel to the target along part of their path and so lie at right angles to the electric field, Fig.5.15. Electrons caught in this crossed field configuration are forced to spiral parallel to the target and so, by increasing electron path length, the chances of ionising collisions with gas atoms are also increased. A region of intense ionisation is created immediately above the target surface which is heavily bombarded and generates a high flux of coating atoms.

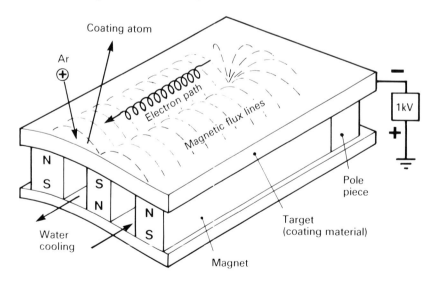

5.15 *Magnetron sputtering source.*

The method can only be used to make magnetic coatings when the target is held above the Curie temperature, otherwise ferromagnetic targets confine the magnetic field and so shield the discharge electrons from its effect. There are numerous variations on this basic theme, differing with respect to biasing, target configuration and magnet; Fig.5.16 shows a rectangular planar source, but cylindrical sources are also available. Magnetron sources coupled with various forms of RF biasing have enormously increased the scope of sputter coating which is nowadays used on a large scale.

5.16 *A planar magnetron source.*

MATERIALS AND APPLICATIONS

Materials

Sputter coating — like ion plating and ARE — can use elemental sources running in an inert gas plasma to deposit elemental films, or elemental sources in a reactive gas to deposit the corresponding compound. Sources that are themselves chemical compounds may also be used, but some change of composition is often found in the coating. The choice between reactive sputter coating from a metal target or coating from a compound is determined by many factors, but both methods are in widespread use as is evidenced by the vast range of materials available as sputtering targets. One catalogue alone lists more than 36 metals, 32 intermetallics and alloy types, 44 oxides, 19 borides, 14 carbides, 17 fluorides, 9 nitrides, 14 silicides, 9 selenides, 10 sulphides and 7 tellurides. It is also possible to make coatings of PTFE and some cross linked polymers. In size and shape the targets range from metal plate several metres long to shaped or planar discs, rings, bars, cylinders, buttons, *etc.* The limits are generally financial rather than technical.

Applications

The following lists are only a sample of the many uses to which sputtered thin films (some only 50Å thickness) are being put.

Engineering
— Corrosion and oxidation resistance, *e.g.* Ni-Cr, M-Cr-Al-Y, polymers.
— Wear resistance, *e.g.* TiN, other nitrides, W, Mo, carbides, borides, diamond type carbon.
— Lubricants, *e.g.* Ag, In, MoS_2, PTFE, selenides, silicides, tellurides.

Electronic
— Metallisation, *e.g.* Al, Cr, Al-Si.
— Resistors, *e.g.* Cr-SiO, Ni-Cr.
— Insulators, *e.g.* SiO, SiO_2, TiO_2, silicides.
— Diffusion barriers, *e.g.* Si(ON), silicides, TiN.
— Semiconductors, *e.g.* GaAs, GaP, In As, In P.

Optical electro-optic
— Lens coatings, *e.g.* MgO, MgF_2.
— Transparent conducting films, *e.g.* In-Sn-Ox.
— Anti-glare coatings, *e.g.* TiO_2-Ag, Cd-Sn-Ox, SnO_2-Cr(N).
— Interference filters, *e.g.* Ge, metal oxides, fluorides, Zn Se-S.
— Luminescents and LEDs, *e.g.* MoO_3, ZnSe.
— Photoconductors, *e.g.* selenides, sulphides, tellurides.

Ion implantation

GENERAL PRINCIPLES

Ion implantation does not produce a coating; the process generates an intense beam of very high energy ions (much higher than in ARE or sputtering) which penetrate into the substrate surface and modify, rather than coat, it (cf carburising). Extensive work in this field has been carried out by Dearnaley and his colleagues at Harwell who have concentrated on beams of ionised nitrogen (boron and carbon have also been explored). Nitrogen ion beams with energies in the range 80-150keV are directed on to the surface of the workpiece and penetrate to depths of $0.1\,\mu m$. The process operates in vacuum (and is therefore a line of sight technique) and the beam is typically of square cross section which expands at an angle of about 20° to cover some 150 × 150mm on the workpiece which may be masked or rotated. Treatment times vary from 1-10hr according to the area to be covered, and the temperature rise may be no more than a few tens of degree centigrade so that fully finished and heat treated parts can be implanted with virtually no change of dimensions or distortion. The only visible change is a slight polishing of the surface.

As the implanted ions are scattered beneath the surface, some of them become trapped in dislocations and other defects thereby effectively pinning them; others form metastable nitrides in some alloys which decompose to give martensitic hardening (unlike gas nitriding of steels which forms stable nitrides). The result of these processes is an increase in hardness and wear resistance, a decrease in friction coefficient and galling tendency, and generation of compressive stresses in the surface which improve both fatigue and corrosion resistance.

A unique and most useful feature of the process is the retention of many of these surface properties after wear to depths that may be 100 times greater than the original implanted depth. It is believed that some of the implanted nitrogen diffuses ahead of the wear front and maintains the original effects.

MATERIALS AND APPLICATIONS

Ion implantation is commercially applied to various steels, to tungsten carbide/cobalt materials, to alloys of titanium, nickel, cobalt, aluminium, copper, to chromium plate, and even to diamond. Extrusion and moulding components for the plastics industry are extensively treated, as are slitter blades for rubber, paper and textiles.

Applications are limited to service temperatures below, say, 250°C for steels and 450°C for carbides, but within these restrictions there are many areas, particularly in metal forming and general engineering, where improvements in service life of a factor of ten are regularly reported for steels, and factors of three or four for carbides and non-ferrous alloys. The marked exception to this latter case is titanium where life improvement factors for prosthetic devices of several hundred are claimed. The cost of treatment is generally in the range of 10-30% of the original component cost.

Chemical vapour deposition

The techniques of chemical vapour deposition (CVD) have been in use for the best part of 100 years, but not specifically for coating purposes; for example, the Mond process for purifying nickel via thermal decomposition of the volatile nickel carbonyl is a CVD process, as are pack aluminising and chromising. As a coating process however, CVD began to play a specialised, but important, role only when the low melting halides of the refractory metals were systematically investigated. Nowadays, the term chemical vapour deposition covers a broad range of processes, all of which use gaseous reagents undergoing chemical reactions near or on the heated surface which is to be coated. The major reactions involved in coating processes are shown in Fig.5.17.

In general, there are three steps in any CVD reaction: firstly, the production of a volatile carrier compound, e.g. nickel carbonyl; secondly, the transport of that gas — without decomposition — to the deposition site; thirdly, the chemical reaction necessary to produce the coating on the substrate. These steps may be quite distinct in both time and space, or they may all occur at the same time within the same reaction chamber — as in pack aluminising

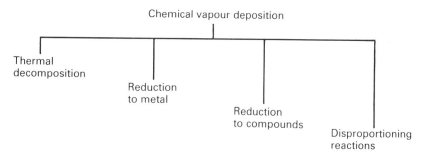

5.17 *Reactions involved in CVD.*

or, indeed, as in the quartz-iodide lamp (in this latter case, the source — the tungsten filament — is being continually recoated by its own evaporated vapour).

The requirements of these three steps impose many limitations on the process, nevertheless, CVD usefully fills a gap where the desired element or compound cannot be satisfactorily deposited by other means to the required thickness. CVD is probably unique in its ability to make pyrolitic carbon and pyrolitic boron nitride.

PROCESS CHARACTERISTICS

The high temperatures and general nature of the reaction involved tend to require less stringent cleaning of substrates before coating them for PVD processes; equally, the high temperatures ensure high density and good adhesion. The coatings are generally rougher than the underlying substrate, but this is partially a consequence of the thicker coatings which range from, say, 10μm up to 1mm. Because the process uses gas at or near atmosphere pressure, CVD has excellent throwing power and coats all exposed surfaces to a high degree of uniformity. This also means that mixed loads of components can be treated, (*i.e.* mixed materials as well as mixed sizes and shapes). The process is particularly well suited to coatings on sintered carbides (see below).

REACTION TYPES

Thermal decomposition

Elemental coatings, chiefly of refractory metals, are produced by simple thermal decomposition of volatile compounds. The most commonly used compounds in this class are: carbonyls, *e.g.* $Ni(CO)_4$; halides, *e.g.* WF_6; hydrides, *e.g.* B_2H_6; and certain organo-metallics, *e.g.* copper acetyl-acetonate.

The carbonyls usually decompose to the metal plus carbon monoxide at temperatures below 200°C, but the substrate must often be much hotter than this if a coherent and adherent coating is to be produced; generally however, substrate temperatures can be kept below 600°C. One severe disadvantage of the carbonyls is their high toxicity.

Halides and hydrides tend to require much higher decomposition temperatures, say above 500°C, but coating requirements again dominate and temperatures above 1000°C are not uncommon. There are many more of these compounds suitable for CVD than there are in the metal carbonyl group.

Organic compounds are not widely used on their own for coating purposes because of side reactions between the products of decomposition, but they decompose at lower temperatures than the halides and are essential in the production of pyrolitic carbon either as a coating or, more importantly, as free standing shapes. Simple hydrocarbons such as the low paraffins are used to make pyrolitic carbon, but substrate temperatures above 1500°C are required.

Reduction to metal

Reduction — usually of halides — by hydrogen is possibly the most important route for the formation of refractory metal coatings; the chlorides tend to dominate this group. Although the presence of hydrogen reduces the reaction temperatures, higher levels are often demanded for dense coatings, *e.g.* 1100°C for titanium, but excellent deposits of molybdenum or tungsten are now routinely produced by hydrogen reduction of the gaseous chlorides at a rate of about 250 μm/hr at 600-700°C. The process is also used to form free standing bodies such as rocket nozzles or crucibles of tungsten and pyrolitic boron nitride, Fig.5.18.

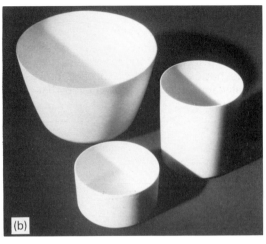

5.18 *Free standing bodies produced by CVD:* a) *Tungsten crucible and lid (courtesy Archer Technicoat);* b) *Pyrolytic boron nitride crucibles (courtesy Fulmer Research Institute).*

Reduction to compounds

A major commercial application of CVD is in the coating of cemented carbide tips on cutting tools, *e.g.* TiC, from the reaction between titanium tetrachloride, methane and hydrogen. Provided that reaction pressures well below atmospheric are used, the reaction temperature can be reduced to 700°C and a TiC layer some 5μm in thickness will give a ten fold increase in tool life. Titanium carbonitride is produced at somewhat higher temperatures but gives a rather longer life. Titanium nitride (from the chloride plus nitrogen and hydrogen) has excellent wear resistance and an attractive golden colour; it is now more widely used than the carbide and is applied to a variety of components ranging from tool tips to watch cases. It is reported that more than 60% of all indexable carbide cutting tools now made in the USA are CVD coated with TiC or TiN (and to a lesser extent) HfN, HfC or Al_2O_3 at a cost ranging from 5-50% of the original tool cost.

In addition to the above, many nitrides, carbides, oxides, silicides and borides are produced as wear resistant coatings by CVD, but the reactants are frequently poisonous and temperatures above 1000°C are usually needed. One exception to this last is that of W_2C which can be produced well below 700°C and may thus have considerable potential for coating heat treated dies, *etc*, without the need for further heat treatment.

Disproportioning reactions

This group comprises a very small number of reactions, possibly the most important being the sequence involved in pack aluminising, other metals treated in this way (*e.g.* germanium, gold) have little interest as engineering coatings in the present context.

PLASMA ACTIVATED CVD (PACVD)

The great improvements in PVD processes brought about by ionising the reactive species prompted a similar approach in CVD where the introduction of a glow discharge further lowers reaction temperatures by many hundreds of degrees Celsius. Whereas CVD of SiO_2 from tetraethoxysilane requires temperatures of some 700°C, the use of RF PACVD reduces the reaction temperatures to 200-300°C and therefore allows quartz films to be deposited on semiconductor materials. Reaction pressures must, however, be reduced below 100Pa to avoid arcing and therefore coating rates are reduced compared with the non-activated reactions and may be as low as 0.3 μm/hr.

Both DC and RF plasmas (with frequencies into the microwave region) have been used, and it has been demonstrated that the quartz reaction above is efficiently activated by UV light. Microwave plasmas are currently of considerable interest for the activated manufacture of diamond films. The films are deposited from a mixture of hydrogen and organic gases containing methyl radicals at growth rates of some 10 μm/hr, but it is reported that Russian workers are producing coatings more than 1mm thickness. Diamond films have a huge potential as wear resistant coatings and in various semiconductor applications, but the first commercial product is in the hi-fi field. The Japanese company Sony has recently (1988) marketed a diamond film high frequency loudspeaker (a tweeter) based on the piezo-electric properties of diamond; the film, being light and stiff, responds particularly well to the high audio frequencies.

MATERIALS AND APPLICATIONS

Metal coatings produced by CVD may be pure and ductile and can be deposited at virtually 100% density. Coatings of, for example, Cr or Ni can be made, but CVD is a more expensive process than electroplating and so is usually applied only to those metals that cannot be electrodeposited. The refractory metals — W, Mo, Re, Nb, Ta, Zr, Hf, *etc*, — and some of their alloys are the major candidates; their coating rates are dependent upon reaction temperature and range from 3µm to 3 mm/hr. The major applications for metals take advantage of the excellent throwing power of the process to produce coatings and free standing shapes at temperatures well below the usual processing levels. Typical products are crucibles, rocket nozzles and other high temperature components, linings for chemical vessels, and coatings for electronic components.

Pyrolytic carbon is remarkably inert and has an excellent thermal conductivity. It is used in chemical vessels, high temperature crucibles and as a coating on nuclear fuels and — in one rather less exotic application — as a bowl liner in tobacco pipes. Pyrolytic boron nitride is chiefly used for crucibles in the manufacture of certain electro-optic materials, and as a coating for cutting tools.

A substantial field for CVD exists as yet another process for producing thin films of TiN, TiC, and other compounds on metal working and forming tools and on components such as impellers, valves, nozzles, *etc*, that are subject to abrasion and corrosion. The widespread application of CVD coatings where very high abrasion resistance is required (*e.g.* thread rolling dies) is largely confined to TiC, but where high lubricity and galling resistance is paramount, TiN is used on, for example, carbide tips, taps, punches, *etc*. To a lesser extent, alumina, which acts both as a thermal barrier and as a protection for the cobalt binder, is applied to carbide tools that are being run at unusually high feed rates.

Although other processes have been applied to the deposition of diamond-like films (i-carbon), the CVD techniques are probably the only ones which can deposit coatings to thicknesses sensibly above a few micrometres. Diamond coatings for wear resistance and for electronic and possibly some optical applications are probably on the brink of a big expansion.

Other compounds that are deposited — particularly by PACVD — for electronic applications include quartz, silicon, silicon nitride and titanium nitride which are used as thin film substrates, dielectrics and insulating layers.

In spite of high coating rates, the generally high temperatures tend to make batch processing times comparable with the slower coating processes; furthermore, the high treatment temperatures often require that alloy steels must undergo further heat treatment after coating. Lower temperatures are possible when the reactions are plasma activated, but PACVD coating rates are generally lower as a consequence of the lower gas pressures that are employed. These and other factors militate against simple comparisons with other processes and, indeed, against a clear cut choice for a specific job, but the CVD processes do have a few outstanding characteristics that make them indispensible for some applications.

Comparison of processes

Although there is a lot of overlap between the characteristics of the various processes, it is difficult to make direct comparisons because so much depends upon the applications involved. Moreover, many of the processes are in an active stage of their evolution and so the suitability of a particular technique for any given application may change dramatically with the advent of new developments. However, it can be said, for instance, that ion implantation is limited almost entirely to wear resistant applications; that CVD is restricted to substrates that can safely be heated to (preferably) 300°C or more; and that thicknesses sufficient for free standing bodies, and thick layers of pyrolytic materials, can be made only by CVD. These statements apart, Fig.5.19 (range of thicknesses), Fig.5.20 (deposition rates) and Table 5.1 demonstrate the wide overlap that exists between the processes for coating thickness, coating rate, coating materials, and the areas of application.

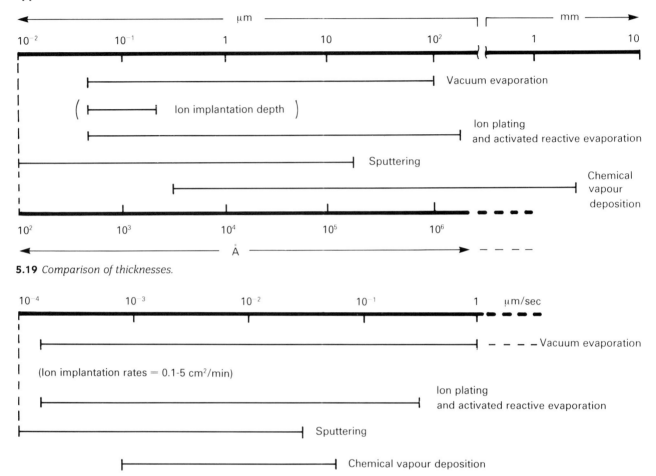

5.19 *Comparison of thicknesses.*

5.20 *Comparison of deposition rates at typical working distances (e.g. sputtering at 50mm, vacuum evaporation at 250mm).*

COATING THICKNESS
The range of thicknesses, Fig.5.19. does not necessarily apply to all materials available to each process, nor to all variants of that process, nor indeed do the values shown necessarily imply that these are the absolute limits, but the charts do show the most commonly encountered ranges for the applications reported. Coating thickness is often determined by the application and not by the process capability.

The above remarks do not apply to ion implantation. Note that Table 5.1 shows not the thickness of ion implantation but the initial implantation depth — the concentration/depth curve being near-Gaussian — however, a unique feature of nitrogen implantation into steels is the inward migration of nitrogen as the surface wears away. Thus, although the major concentration of implant is initially confined to a thin subsurface layer of only a fraction of one micron, the wear resistance persists to a much greater depth.

COATING RATE
Comparison of coating rates for a given material is complicated by the fact that different processes use different source to substrate distances. With the exception of ion implantation, the coating vapour stream issues as a cone, or even totally fills the reaction chamber, thus coating rates vary in a complex manner with coating distance; Fig.5.20 shows the range of rates usually found in practice. Virtually all process rates can be increased by a multiplication of sources and — to a lesser extent — by an increase in the power input, but the coating quality may suffer if too high a rate is attempted.

The magnetron sputter source is expected to continue to increase in efficiency as a result of the intense developments that are taking place. There has been such an increase in magnetron sputtering rates that the process has become commercially competitive with evaporation techniques for many materials, and there is as yet no indication that the limit has been reached, particularly for area coverage.

Ion implantation rates cannot be quoted in the same way as coating rates. The total dose of implant is the important factor (usually 3×10^{17} nitrogen ions per square centimetre) and therefore the treatment rate is the time taken to implant unit area with the required dose. Treatment rates of up to 5 cm^2/min of nitrogen implant are currently possible and, again, there is no indication that the limit has been reached.

COST

Direct comparisons of coating cost are not generally possible, but broad generalisations can be made. Vacuum evaporation by resistance heating is probably the cheapest in all respects, but the introduction of electron bombardment heating or RF power can significantly affect the capital and running costs. Ion plating and its associated techniques require fairly expensive equipment in comparison with vacuum evaporation because there are so many more factors to control and because coating efficiencies are lower.

The capital cost of sputtering equipment is high and its complexity is increasing, but it is undergoing continuous development (thanks to the electronic industry) and basic unit costs are not increasing too rapidly in real terms. It is too early to make cost comparisons between ion implantation and the other methods because the process has only recently been launched on a commercial scale; area treatment direct costs of ion implantation are modest but capital costs are high. However, here again development (principally for machine tools) is continuing and is bearing fruit.

CVD is a remarkably inexpensive method of forming some of the refractory metals when compared with the traditional techniques, and even its unique ability to produce pyrolytic materials is not an inherently costly process, but the indirect costs caused by the toxicity of many of the compounds used can add significantly to the capital and running costs.

GENERAL COMMENTS

In conclusion, apart from those examples already mentioned, the areas of dominance for any given process are not well defined. Doubtless in time the picture will become clearer, as experience is built up for the various applications, but there will always be large areas of overlap where the superiority of any one particular method will be determined by specific factors.

Design for physical and chemical vapour deposition

GENERAL CONSIDERATIONS

Physical vapour deposition techniques are many and varied and include warm or hot treatment in vacuum whereas CVD employs at least two highly reactive gases which generate the coating by high temperature reaction either in the gas phase or on the substrate. The various processes have been considered in detail but some basic considerations are important for the designer.

The whole component whose surface (or part thereof) is to form the substrate must be compatible with the process. Thus, the substrate temperature for PVD will nominally be at ambient or somewhat above but temperatures in excess of about 500°C are rarely experienced by the part. CVD processes, on the other hand, tend to operate at much higher temperatures that will generally be in the range 500-1200°C, but may be much higher. The vacuum requirements of PVD, and the high operating temperatures of CVD, demand that the component being treated must not contain materials which have high vapour pressures under the deposition conditions, e.g. brazes containing Cd or Zn. Plastics, however, can be coated under certain conditions. Both CVD and reactive PVD techniques use reactive gases, therefore precautions must be taken against leakage of these gases on to uncoated areas. Articles to be coated must allow free access both for evacuation and for deposition, *e.g.* recesses may not be easily evacuated, or they may present difficulties during glow discharge cleaning, or they may not be sufficiently accessible to the coating vapours. In general, the design requirements that apply to, for example, electroplating, apply equally to PVD and CVD coating.

DESIGN REQUIREMENTS

1. The design should not contain deep, blind holes.
2. Small bores are difficult to coat. The component preferably should be designed so that it can be split.
3. Holes or recesses with a width to depth ratio of 1:1 can be coated but a broader ratio is preferable.
4. There should be no surface voids and the component should not be porous or contain fissures, *etc.*
5. There must be no burrs on the edges of the surface to be coated.
6. Parts must be assembled, *e.g.* screwed together, or shrink fitted.
7. Large variations in cross section can give rise to temperature variations during the process and should be avoided if possible.
8. Masking must be mechanical and, because fit-ups must be tight, it entails careful design and probably consultation with the sub-contractor.
9. Some means of fixturing may be necessary and should be discussed with the sub-contractor.

SURFACE CONDITION

1. The surface of the parts must be bright and untreated.
2. Suitable surface finishes are, for example, ground, polished, fine finish spark eroded, or spray lapped. Blunt or worn grinding discs must not be used because cold laps must be avoided.
3. Polishing agents should be removed using an appropriate solvent.
4. Parts should be lightly oiled or covered with a sealant as soon as possible after machining or cleaning to protect against corrosion.
5. Parts must be free from machining chips and foreign particles, especially in blind holes.
6. The surface to be coated must be free of any coating or surface treatment such as electroplating, carburising, nitriding, chromating or aluminising.
7. The surface must not be burnished.
8. There must be no rust, paint or colour identification markings.

DATA TO BE SUPPLIED TO THE PROCESSOR

1. Type of material and specification if available.
2. Tempering temperature and any heat treatment to which the part has been subjected.
3. Drawing showing dimensions and indicating clearly the surfaces to be coated, surfaces which may be coated, and any surfaces where coating is to be avoided.
4. If polished, details of the polishing process.
5. For machine parts, type of tool or description of its use.
6. Hardness, indicating the point where measured.
7. Identification marks/numbers to be retained.

Plastics coatings

The use of plastics materials for engineering coatings developed from the adaptation of thermal spraying torches, originally designed for spraying metal powders and patented in the 1920s, to handle plastics which started to become available in powder form in the early 1940s.

The useful properties of many plastics have led to development of thermal spraying equipment designed specifically for spraying them in powder form. In addition, other techniques, including electrostatic spraying and the use of fluidised beds have been adapted to apply coatings. The various possible combinations of material and application process that have emerged have led to a wide range of uses for engineering and decorative purposes.

The majority of coatings are applied for protection to the surface of metallic substrates, often allowing use of cheaper construction materials. In addition, plastics coatings have been applied to substrates such as concrete for waterproofing and to decorative gypsum tiles for improved appearance and waterproofing.

While such coatings have an upper service temperature limit of about 250°C or less, depending on the plastic used, all have good resistance to corrosion by a wide range of media, are tough and have insulation properties. Hardness and wear resistance depend on the material type.

Plastics coatings are widely used for their corrosion resistance, either as a substitute for cathodic protection by zinc or aluminium, as an overlayer to these to extend life, or to provide a durable and decorative finish by using one of the many colours in which most plastics are available.

When used for protection against corrosion, plastics coatings offer significant benefits over paints in terms of ease of application and durability in service. In addition, plastics can provide superior resistance to abrasion and a wider usable temperature range. Coatings are available suitable for use in bearing applications and to meet the hygiene requirements of handling foodstuffs.

Application of plastics powders to a component requires the presence of sufficient heat to fuse the powder. As this is only a few hundred degrees Celsius, there is little or no problem with metallurgical changes in the substrate and provided that the metal surface is correctly prepared, adhesion of the coating is excellent.

Materials

A selection of materials used in industry is listed in Table 6.1. The majority are thermoplastics and offer a range of properties with an upper service temperature of about 250°C for polyether ether ketone. For higher temperatures, a thermosetting epoxy resin is used. This type of material requires a post-coating curing treatment unless a curing agent is added to the powder.

Particle size of the plastics powder is important. The size for electrostatic spraying (30-60μm) is too fine for thermal spraying as the particles would overheat in the flame; fluidised beds and flame spraying use a coarser powder of 80-200μm.

A common material is polyamide (Nylon) which has been in industrial use for about 30 years. Nylon 11 is the most widely used polyamide for coating purposes. In addition to characteristics listed in Table 6.1, it has a well established use with foodstuffs. Powder is available in a variety of colours.

The ethylene/vinyl acetate copolymer (EVA) is also widely used for coatings. In addition to the properties listed, it has a wide processing temperature range of 180-350°C and a relatively low cost.

Thermoplastics coatings are suitable for local repair without removal of the original coating.

Table 6.1 Plastics coatings materials

Material	Melting temperature, °C	Application techniques			Properties	Workpiece preheat temperature, °C
		Thermal spray	Electrostatic spray	Fluidised bed		
Thermoplastics						
Ethylene/vinyl acetate copolymer (EVA)	108	√	√	√	Fairly soft, resilient, tough at low temperatures, high resistance to chemicals and UV. Wide process temperature range. Included in BS 5977 Pt 2.	250
Low density polyethylene	111	√	√	√	Cheap, fairly soft, resists chemical attack; insulator	200
High density polyethylene		√	√	√	As above but tougher, more heat resistant, surface less glossy	200
Polypropylene	120	√	√	√	Relatively soft, good resistance to chemical attack	150
Polyamide/ polyether copolymer	128	×	×	√	Flexible, impact resistant and chemical resistant	150
Macromolecular thermoplastic polyester	174	√	√	×	Hard and resists impact and chemicals at elevated temperatures	250
Polyamide (Nylon 11)	186	√	√	√	Hard, resists wear and impact and chemicals at elevated temperatures except concentrated mineral acids	220
Polyether ether ketone	334	×			Tough, resists wear, chemicals and service temperatures up to 250°C	350
Polyvinyl chloride (PVC)	350	×	×	√	Tough, resists chemicals including mineral and vegetable oils. Good insulator	
Thermosets						
Epoxy resins	—	×	√	√	Usable at higher temperatures than possible with thermoplastics; requires curing to develop properties	150

Use of other plastics for coatings is under development and specific applications are reported from time to time in the technical press. Experimentation needs to be undertaken with caution as the temperatures involved in applying the coatings may lead to the emission of toxic fumes from some plastics.

In addition to being used as straight polymers, an improvement in resistance in abrasive wear has been obtained by addition of mineral substances such as glass flake to the polymer powder.

Application processes

FLAME SPRAYING
Spraying plastics developed in the 1950s using modified equipment originally developed for spraying metals, but as the requirements of the materials are different, guns now in use have been designed specifically to handle plastics.

A build-up of hot plastic inside the gun caused by the flame temperature must be avoided by surrounding the powder jet by cooling air, Fig.6.1. and 6.2.

The flame is produced by combustion of a fuel gas such as acetylene, hydrogen or propane in oxygen or air. It is necessary to ensure that the particles of plastic are raised to their melting point but do not suffer any degradation in properties caused by overheating.

Powder is fed to the gun either from a hopper mounted directly on the gun itself or conveyed by air in a hose from a separate powder feeder. The latter method has advantages for work with long spraying times or that is mechanised.

6.1 *Magnum torch for plastics coating, designed for touch-up work on damaged components and for site work. The powder consumable is held in the torch mounted hopper (courtesy Croboride Ltd).*

6.2 *Naval exhaust muffler being thermally sprayed with Nylon (courtesy Croboride Ltd).*

Advantages
1. A portable and versatile process suitable for workshop or site work.
2. Thick coatings up to over 1mm are possible.
3. Low capital equipment cost.
4. Rapid changes of powder are possible.
5. Can be mechanised or manually operated.

Disadvantages
1. Coating uniformity and quality is dependent on manual operator's skill.
2. Overspray losses may be high on small or narrow workpieces.
3. Surface finish may be inferior to other processes.
4. Line of sight process; access limitations.

Selection of correct spraying parameters and their control is essential to production of deposits of the correct and repeatable quality.

FLUIDISED BED COATING
An early technique consisted of preheating the component to be coated and then rolling or dipping it in plastic powder causing the powder in contact with the workpiece to melt, building up a coating which, with further heating developed a smooth, well bonded continuous film. However, in the 1950s, the technique was considerably enhanced by adapting the fluidised bed technique which provided improved coverage especially of irregular surfaces, and improved productivity.

Advantages
1. Improved uniformity of coverage and coverage of irregular and internal surfaces.
2. No powder losses of the type associated with spraying.
3. Simultaneous coating of external and internal surfaces.
4. Better temperature control of workpiece possible as coating is instantaneous all over.
5. Thick coatings possible in a short time.
6. Process easily mechanised.

Disadvantages
1. Essentially a workshop process, not portable.
2. Higher capital cost of equipment than thermal spraying.

3 Large volume of powder held in fluidised bed.

4 Quick colour changes not possible.

5 Masking difficult because of preheat temperature required.

6 Size of workpiece is controlled by size of fluidised bed and preheating oven.

ELECTROSTATIC SPRAYING

In this process, the powder delivered from the spray gun is electrostatically charged and propelled at low velocity by air or a revolving spray head to the workpiece which is 'earthed'. Care is necessary at this stage as too much air pressure causes the powder to bounce off the substrate and makes recesses difficult to coat.

The process produces a coating thickness which is self-limiting, as the increasing thickness of coating increases the electrical insulation of the surface of the substrate. This effect can be overcome by preheating the substrate and applying several coats. After spraying the component is transferred to an oven to cure the coating and to develop a continuous film.

A major advantage of electrostatic spraying is the wrap-around achieved. The external surface of a 25mm diameter tube, for example, would be completely covered during the spraying process even if the tube were not rotated during spraying and the gun was only traversed along its length. Manufacturers of quality tubular articles, such as furniture, bicycles, *etc*, are major users of the process.

Advantages

1 Good wrap-around without workpiece positioning is possible.

2 Process can easily be mechanised.

3 Thin films are possible (50μm).

4 Good edge cover.

5 Temperature control of oven ensures correct fusing of coating.

6 Masking easier than with other processes.

Disadvantages

1 High capital cost of plant.

2 Equipment not portable — a workshop process.

3 Size of workpiece is controlled by the capacity of the equipment.

OTHER TECHNIQUES

Small parts

Parts which are too small to be handled individually are heated and dropped into a vibrating or tumbling bed of powder. The parts are later postheated to produce a pore free film. Some systems allow for coating without suspension points or touch marks. Up to 50 000 parts per hour are possible using unskilled labour.

Vessel interiors

For components such as fire extinguishers and hot water cylinders, a system exists for introducing a measured quantity of powder into the preheated part and then moving the part in a planetary fashion to distribute the powder evenly on the inner surface while heat applied from outside fuses it in position.

Pipe interiors

To coat interior surfaces of complex parts, a low vacuum is applied to pull the powder through the preheated component. Extremely complex pieces are easily coated by this system which has been used for lengths up to ten metres and diameters of one metre. Heat exchangers have also been coated in a similar manner.

Continuous wire coating

In continuous wire coating, wire is drawn through a fluidised bed of powder and heated in it by an induction coil, thereby picking up a fused coating. A second coil above the powder flows out the coating.

Continuous tube coating

Continuous tube coating is similar to the wire coating process, but the powder is applied in a 'cloud chamber' and the hot tube is cooled in water before further processing. Continuous production is possible at high speeds with a closely controlled thickness of coating.

Flock spraying

Components which are too large or heavy to handle in a fluidised bed are heated in an oven and then sprayed by a simple venturi type gun, with or without electrostatic charge. Powder fuses to form coatings up to 1250μm (0.05in) thickness if required, which are then restored if necessary to complete the fusion.

Designing for plastics coatings

SURFACE PREPARATION

Coatings of plastics on metallic substrates must be applied under carefully controlled conditions. It is essential that the substrate is perfectly clean before any coating operation commences. This means that all traces of scale, rust, grease, paint or other contamination must be removed. Failure to do so will result in loss of adhesion between the metal and the coating.

The cleaned surface is normally blasted using angular iron grit. In addition to the cleaning effect, this roughens the surface and provides a mechanical key for the coating. The surface after blasting is extremely active and will oxidise rapidly, especially in damp air, so coating must commence immediately after blasting. Grit used is to SA2½ standard, G17. For thin film coatings where the roughening effect of G17 grit would be too great, use of G7 grit is recommended.

As an alternative to grit blasting, chemical cleaning followed by phosphating or chromating is used, especially for thin coatings obtained by electrostatic deposition. Phosphate coatings are unsuited to the high temperatures used for fluidised bed deposition. Specialist advice is recommended when choosing such a treatment.

Although grit blasting is the best preparation for all thick coatings, use of a primer system before applying the coating is advisable in service applications. These improve the adhesion of the coating to the substrate and are usually recommended for electrostatic applications of Nylon. Primers based on epoxy phenolic formulations are applied by conventional spray guns to a thickness of about 7-12μm as a dry film. Spraying can be carried out with electrostatic assistance if required and will dry fairly quickly. The temperatures involved in thermal spraying may damage the primer and this may preclude their use with this process.

Thin components can easily be distorted by abrasive blasting — chemical cleaning and use of primers can be the answer.

COMPONENT DESIGN

Surfaces to be coated must be accessible not only for the coating process chosen but also for the surface preparation. Because of the heating required to fuse coatings of plastics, it is necessary to provide vents in any hollow sections; cleaning fluids will enter through these and they should therefore be sited so that the fluids can be completely drained from the cavities. Subsequent sealing of these vents may be necessary to prevent internal corrosion of the unprotected surface during service.

It is good practice to avoid sharp corners on parts to be sprayed. The coating processes vary in their ability to coat these as uniformly as other surfaces, but the coating may be thinner on sharp corners and will be more vulnerable to damage in handling, especially if the part is large or heavy.

Summary

Given the wide range of industrial wear problems, plastics materials occupy a useful niche in selection of cost effective coatings available to the design engineer.

As with materials and processes covered in other chapters, coatings can be applied in the factory as part of the original design of the component, or in the field, the latter may be to repair a part which was plastics coated when new, or to extend the life of an uncoated part. Depending on circumstances, manual or mechanised versions of the processes can be adopted. However, strict adherence to proven operating parameters is essential to satisfactory and predictable service life of the part. Where manual application is envisaged, use of properly trained operators is important.

Points requiring special care are correct design and preparation of the part, control of pre- and post-heating temperatures and avoidance of moisture condensation on parts heated by direct flame.

Finishing of surface coatings applied by welding and thermal spraying

CHAPTER 7

It will be clear from previous chapters that surface coatings vary widely in their characteristics. Some are used in applications where they can be put into service in the as-coated condition, but many are applied to engineering components which demand a dimensional accuracy and surface finish which cannot be met without a precision finishing operation.

Coatings designed to resist wear are, by definition, likely to be more difficult to machine than ordinary construction materials and therefore call for different techniques. This chapter is devoted to the various finishing processes used for achieving dimensional accuracy and surface finish on coatings applied by weld deposition and thermal spraying. Finishing of electrodeposited coatings is dealt with in Chapter 4; coatings applied by PVD and CVD in Chapter 5.

Fused and welded coatings

By their nature, fused and welded hardfacings are of an irregular form and generally need to be machine finished. The purpose of this section is to discuss the various methods used in machining cobalt and nickel based alloy deposits, and to set some reference points from which to work, but it should be emphasised that the parameters given are only intended as a guide.

THE NEED FOR MACHINING

Inevitably, machining operations performed on hardfacing materials are expensive because of the wear resistant nature of the alloys, so consideration should be given to the reasons why machining is necessary. The most common requirements for machining involve the need to produce:

1. Closer dimensional tolerances than a deposited finish can achieve;
2. Better surface finish than may be expected from deposition;
3. Geometric shapes which deposition methods cannot achieve.

The above need to be considered, not only when selecting the method of machining, but also in the original design of the component. The design engineer will need to know which alloys may be selected to enable dimensional tolerances and surface finishes to be achieved and must be aware of the relative costs of the procedures to be used in achieving those requirements, if cost is a major factor to be considered within the design parameters. As the required degree of precision increases, so does the cost. This is even more evident when dealing with wear resistant coatings. The relationship is not a linear one, as it costs little, if any, more to make a part within 0.3mm of its nominal size, than to within 0.5mm. However, it costs much more to produce a part to within 0.003mm than to within 0.03mm of nominal size.

When considering the production economy of machining wear resistant deposits, it is necessary to examine the complete system. Achieving dimensional tolerances and surface finishes depends upon a combination of:

1. The machine tool and its condition;
2. The degree of difficulty in machining the material;
3. The type of cutting tool required.

It is for these reasons that the cutting parameters suggested should be regarded only as 'bench marks', rather than the ultimate achievable.

Before considering machining parameters in detail, time should be given to understanding what may be achieved in terms of dimensional tolerances and surface finishes. These are set out in Fig.7.1. The relationship between surface finish and relative production times for machining these materials is shown in Fig.7.2.

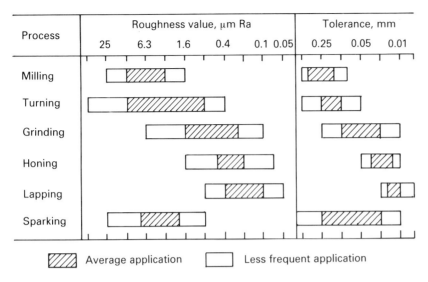
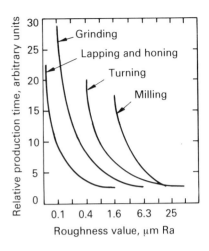

7.1 Typical surface finishes and dimensional tolerances produced by various processes on hardfaced deposits.

7.2 Typical relationship between surface finish and relative production time for machining hardfaced deposits.

TURNING

Most hardfacing alloys below 60RC can be turned. However, the set up should be rigid so that the tool is presented to the work without any deflection or flutter. Slides should be tight, tool projection at a minimum and the tool section as heavy as practicable. For most applications a negative rake is preferred.

Knowing the condition and capability of each machine tool is essential if standards are to be maintained. Regular monitoring of machining operations, by comparisons against set standards, plays a useful role in avoiding excess costs whilst achieving the demanded quality standards. Deterioration of the machine tool will be detected, indicating that attention is required, or even that the machine tool is no longer economically viable for a particular operation or standard. In the selection of feeds and speeds, the three main considerations are:

1 Amount of stock removal;

2 Surface finish requirements;

3 Dimensional tolerance requirements.

If the turning operation is to finish the diameter in question to size, it is usually necessary to take a roughing cut, followed by a finishing cut to size taking surface finish requirements into account. The roughing operation would normally be carried out using the combination of feeds, speeds and depths of cut to give best economic stock removal, but for the finishing cut other considerations must be made. Because these materials are resistant to wear, it follows that there is a natural resistance to the turning tool. This has the effect of 'tool push off', making it almost impossible to pick up a cut within less than 0.1mm, where a tool may have failed. It is therefore essential, in making the finishing cut, to select a range of speeds and feeds which will guarantee a tool life at least as long as the cut will take, and that at the end of the cut, the tool will still be performing well enough to give the required surface finish. It is this feature of push off that makes final size prediction uncertain and close tolerance work difficult. Where fine feeds are required it normally follows that increases in surface speeds may be used. Similarly, where coarser feed rates are advantageous in cutting some of the work hardening materials, then surface speeds should be reduced.

Three types of cutting tip are recommended:

1 Tungsten carbide in the K10 range ISO;

2 Ceramic — mixed;

3 Polycrystalline CBN.

Selection depends upon the grade of material and the soundness or consistency of that material. Typical speeds and feeds which may be used are shown in Table 7.1. Hardfacings with a hardness value of less than 60RC

Table 7.1 Guidelines for machining

Type of alloy	Surface speed, m/min	Depth of cut, mm	Feed rate, mm/rev
Tungsten carbide			
Cobalt based	7-25	<5	0.1-0.4
Nickel based	5-30	<3	0.2-0.3
Ceramics			
Cobalt based	36-75	<4	0.2-0.4
Nickel based	30-80	<2.5	0.3-0.4
Polycrystalline CBN			
Cobalt based	170-240	<4	0.2-0.3
Nickel based	80-120	<3	0.2-0.3

can be easily machined using tungsten carbide tooling as described, however, variations in both chemical composition and hardness values make machining with ceramics or polycrystalline CBN less predictable. Interrupted cuts, such as may occur in uneven coatings or with imperfections in the overlayed material, can cause catastrophic failure of the ceramic cutting tool whilst other variations as yet more difficult to understand, but possibly an effect of base material dilution, tend to affect the machining parameters when using polycrystalline CBN. Where results have been obtained, attention to detail in the deposition and heat treatment operations to give repeatability at the subsequent machining operations is essential. Many examples are available to indicate that most nickel and cobalt based deposits can be machined using polycrystalline CBN or ceramic tips, but it is advisable to record successes for future reference.

MILLING

Milling is restricted to the softer hardfacing alloys, and the cutting tools normally used are tungsten carbide. The nature of the milling operation is one of a series of interrupted cuts and ceramic tips cannot usually withstand these forces generated in the milling of wear resistant coatings, whilst polycrystalline CBN tools are not in contact with the workpiece consistently enough to achieve the desired cutting parameters. There will be, however, instances where both ceramics and polycrystalline CBN tools are effective. For the purpose of this chapter, only tungsten carbide tools are considered.

The three areas of machining where milling is an effective operation are:

1 Large flat areas;

2 Machining angles;

3 Machining grooves (*i.e.* keyways and grease grooves).

A vertical milling machine is considered to be advisable because of the rigid set ups that are required to achieve the best results. Milling cutters should be selected which use indexable inserts of the negative geometic form (SNUN, TNUN, RNUN, *etc*) as this ensures maximum strength at the cutting edge. The size of the cutter selected should be such that the maximum length of cutting stroke is achieved, the best results being obtained in cutting a groove in the material (180° arc of contact). Surface speed at the cutting tip should be approximately 60% of that used for turning with feed rates of around 0.05mm per tip per revolution. Depth of cut should be no more than 3mm, although this may be increased for some of the softer alloys. The wear pattern, flank or crater, on the insert will indicate where the rpm or feed rates should be adjusted. When machining grooves or angles, roughing and finishing operations should normally be used if stepping or breakout is to be avoided on the sides of the grooves or along the angles being machined.

For example, for face milling a block 180mm wide overlaid with cobalt based deposit, hardness 36RC, use: milling cutter 200mm diameter negative geometry with 14 inserts; cutter speed 20rpm; depth of cut 2mm; feed rate 14 mm/min.

DRILLING

Production drilling of holes in most hardfacing alloys presents few problems. If close dimensional tolerances are required, then the set up must be rigid. Breakout can occur if drilling through holes, but this may be overcome by using progressively larger drills, and finally drilling into the deposit from either side.

The least expensive way to drill holes of up to 20mm diameter in hardfacing materials is by using standard masonry drills. These usually cut well oversize and the clearance helps to avoid overheating. Speeds of around 400rpm with a feed rate of 0.01 mm/rev are usual when drilling holes of up to 7mm diameter whilst for holes between 8-20mm diameter speeds of around 170rpm are usual.

It must be remembered that usually it is only necessary to drill through the deposit into the base material using the techniques described above. From the point where the drill meets the base material, conventional drilling techniques, with regard to that material, should be used. Coolant should always be used as it helps to avoid cracking and gives greater life to the cutting edge

GRINDING

All fused and welded hardface deposits can be ground to give a surface finish of $0.4\mu m$ Ra. For small batch production, wheels normally used for grinding high speed steel can be employed provided they are kept open and free cutting, and that a copious supply of coolant is used.

Cylindrical grinding

Before finish grinding, it is advisable to turn the deposit to within 0.4-0.6mm of finished size, if possible. For most grinding operations, aluminium oxide grit wheels in the range A60J with vitrified bonds can be used. Approximate peripheral speeds of around 25 m/sec with a workspeed of around 15 m/min is a good starting point, but with some of the harder deposits it may be necessary to reduce the wheel speed if possible, or if this is not possible, the work speed to minimise the possibility of cracking.

Wheels using CBN normally give greater stock removal rates when grinding harder deposits, but precise bands of feeds and speeds will need to be selected to give the best surface finishes. For internal cylindrical grinding, the same parameters are generally used but the rigidity of the grinding quill has an important part to play in achieving surface finish and good stock removal.

Surface grinding

Most surface grinding of hardfaced deposits is carried out on vertical spindle, rotary table grinding machines, using ring wheels or segmental wheels if economic stock removal is to be achieved.

Selection of wheel types depends upon:

1 The surface area of deposit to be ground;

2 The type of hardfacing alloy;

3 If steel and deposit are being ground together.

Selection guidelines are as seen in Table 7.2.

As with other types of grinding copious supplies of coolant are required.

Using conventional surface grinding techniques (horizontal spindle reciprocating table machines), wheels in the A60J to A120K range are recommended, but advantages in stock removal rates may be found on the harder alloys by using CBN or diamond grinding wheels.

HONING

Honing is recommended as a reasonably fast and accurate alternative to grinding when finishing deposited bores where either dimensional tolerances are close or where a fine surface finish is required. Because most of the bores deposited will be more than 75mm diameter, the type of honing machine used will normally be either the beam hone or large vertical hone. The component should have been machined to within 0.1-0.2mm of finished size. Experience suggests that CBN stones give best results where surface finish requirements are no better than $0.1\mu m$ Ra, but for general use stones in FIF3D (Delapena) give satisfactory performance.

Table 7.2 General recommendations for grinding wheels for deposits of cobalt base alloys of Group 3

Purpose	Approximate peripheral speed, m/sec	Recommended wheels
Peripheral wheels		
Surface grinding	27.5	A60JV
Form grinding		
(use finer grits for sharper forms)	27.5	A100GV A120KV A150GV
Ring wheels		
Mixed surfacing alloy and steel	19	A30DZB
Surfacing alloy only		A50BB
Segmental wheels		
Mixed surfacing alloy and steel	19	A36FB
Surfacing alloy only		A46EB
Internal grinding		
Bores over about 100mm diameter	25	A54JV A54MV
External grinding		
For wheels up to 305mm	27.5	A60JV
For wheels over 305mm		A46JV

Note: Wheels carrying the same marking from different manufacturers do not necessarily have the same cutting action. Check that makers' maximum safe speed is not exceeded. The wheel descriptions above are based on the recommendations of ANSI B74.13.1970 and BS 4491

LAPPING

Lapping is an operation where the complete combination of machine, material and type of abrasive needs to be considered as a total system. Coarse diamond or silicon carbide grit, together with pressure, gives best stock removal rates if surface finish requirements are not too fine. Fine diamond powder (3-5μm) or aluminium oxide powder (520 grit) is necessary, however, if surface finish requirements are better than 0.2μm Ra. Very fine surface finishes take time to achieve, therefore a roughing operation can be advantageous before fine finishing, however two machines will be required as to change a machine from coarse to fine grit a complete wash down is necessary.

SPARK EROSION

Little needs to be said about machining by spark erosion, but it should be stated that, as with most other processes, quality of finish versus production time is the main consideration. The operation itself is one which causes few problems whether using the wire erosion technique or the more conventional copper or carbon electrode method. Imperfections in the deposit, however, when using wire eroding can have a dramatic effect on quality and may cause wire wander.

Thermally sprayed coatings

For turning softer coatings established tooling, such as high speed steels and sintered carbides, gives the best results. The biggest problem, often, is to ensure that porosity in the deposit is not sealed by surface deformation.

Successful machining starts at the design stage, where adequate support must be given to the ends of a deposit and allowance made for elimination of any ragged edge from which cracking may propagate. Correct substrate preparation must be employed with, if necessary, use of bond coats and/or threading to improve coating adhesion. Selection of material, too, is important. It may, for example, be advisable to spray on a thin coating of a softer alloy on to a hard deposit to facilitate final machining. It is important to select the surfacing material with consideration of the machining equipment available.

While a very low carbon steel provides good wear properties under lubricated conditions, increasing the hardness, which may be essential, decreases machinability and requires better tool tips. As the thickness of the deposit increases it becomes more desirable to select materials with low shrink characteristics, to reduce internal stresses which can cause cracking as the additional machining stresses are superimposed. Various bronze alloys are in regular use. If a low cost commercial bronze is satisfactory this may be used but it will not machine to as good a finish as an aluminium bronze. The latter shows excellent cohesion to the substrate, a coherent dense structure and very good wear resistance. A good finish is more difficult with sprayed nickel than sprayed Monel although the resistance needed to some corrodants may dictate the choice of nickel.

Selection of spraying process may affect machinability but whatever the process it is important that it is carried out correctly. Lack of control may lead to abnormal particle size, oxidation or other chemical effects, excessive porosity or loss of cohesive strength, any of which can cause deterioration in the quality of the machined surface and possibly porosity or cracking.

Sufficient working data have been established for machining and grinding of most commercially developed materials used for metal spraying. Some provide better finishes than others but commercial finishes within commercial tolerances can be achieved with all. However, the machinist should not use a tool tip tailored for cutting cast or wrought metal nor should the grinder select wheel grades or grit sizes normal for these materials.

TURNING

All but the hardest deposits can be machined with high speed tooling but faster cutting rates are obtainable using cemented carbide tool tips as indicated in Table 7.3. Such tooling is not normally used for softer materials. Carbide tips provide better finishes as higher speeds may be used and some of the harder alloys, which otherwise require grinding, may be machined easily. The low coefficient of friction and high resistance to welding enables carbide to cut freely with much less rake and clearance than is needed for ordinary tool steel, again assisting in providing a better finish. The oxides and carbides frequently encountered in sprayed coatings subject the nose of the cutting tool to much abrasive wear and the hardness and wear resistance of carbides again provides an advantage over ordinary tool steels. With harder deposits, use of the latter will lead to problems in maintaining dimensions.

Table 7.3 Typical basic turning parameters for sprayed coatings

Metal	High speed steel tools		Sintered carbide tools			
	Surface speed m/min	Feed, µm/rev	Surface speed, m/min		Feed, µm/rev	
			Roughing	Finishing	Roughing	Finishing
Pb, Sn, Babbitts	45-75	125-175	—	—	—	—
Al, Cu, brasses, bronzes	30-40	75-125	75-90	90-105	150	75
Nickel alloys	30-40	75-125	60-75	75-90	100	50
Low carbon steels	23-30	75-125	23-30	23-30	150	75
Medium carbon steels	15-23	75-125	15-23	15-23	100	75
High carbon steels	—	—	10-12	10-12	100	75
Martensitic stainless steel	—	—	10-12	10-12	100	75
Austenitic stainless steel	30-40	75-125	30-40	40-45	150	75

Sprayed metal coatings are porous. For ordinary build-up work on machine components used in lubricated conditions development of self-lubrication is advantageous as the oil tends to seal the pores. Thick coatings are relatively impervious but with thin coatings, or parts used in active corrosion environments or under high pressure, such as hydraulic press rams, sealing is necessary. There can be advantages in sealing shafts and other machine parts even if they are not used under high pressure or in corrosive conditions. Self-sealing coatings of tin, copper, aluminium or 18/8 stainless steel may be sealed by using high pressures and blunt tools, by rolling, or by peening. Such methods may be used to control surface finish and, at times, dimensional tolerances.

GRINDING

Grinding is often the preferred finishing method for the harder surfacing materials and for pure metals such as Cr, Nb, Ta and W. The grinding wheels for sprayed coatings should be relatively coarse grained, of open structure and low bond strength. Silicon carbide is frequently used except for deposits containing high concentrations of tungsten carbide. General purpose wheels approximate to the following specification: grit size 60; hardness medium soft; structure open; bond vitrified and for a silicon carbide wheel would have a typical BSI marking C60-K10-V, to which the manufacturer's prefix and suffix may be added.

The wheel specification may be changed depending on the particular job. Thus a softer wheel should be used if there is a large area of contact, if faster stock removal is required or if a wide wheel is to be employed with light pressure, while the hardness should be increased to provide a finer finish or if the area of contact is small and a narrow wheel is used with high pressure. As sprayed metals tend to 'load' the wheel a relatively coarse grain and a low bond strength are needed so that the surface breaks down with comparative ease. A dull or glazed wheel or use of an unsuitable abrasive can exert a high shearing force generating much local overheating, which is aggravated by the comparatively poor thermal conductivity of the coating, leading to burning and heat checking.

Wet grinding is preferable to dry grinding and should be used whenever suitable equipment is available. It presents few problems provided the right wheels are used and the following precautions observed: always use light pressures; avoid glazing of wheels; use finer grits for finer finishes; use copious quantities of coolant, see Table 7.4. In dry grinding use substantial equipment to prevent excessive wheel vibrations, see Table 7.5.

Table 7.4 Typical wet grinding parameters

For all processes	*Wheel speed 1650-2000 smpm* *Work speed, smpm R15-21: F21-30*
Centreless grinding	*Regulating wheel — angle°, R 2-4; F 1-2; rpm R 15-30: F 20-40*
Internal grinding	*Table speed, m/min 1.25-1.8*
Cylindrical grinding	*Fraction of wheel width advance per rev. R $\frac{1}{4}$ - $\frac{1}{2}$: F $\frac{1}{12}$ - $\frac{1}{6}$*
Surfacing grinding	*Table speed, m/min 12-30: Cross feed, mm R 0.15-0.30: F 0.04-0.08*

smpm = surface metres/min
R = roughing, F = finishing

Table 7.5 Typical dry grinding parameters

	Wheel speed, smpm	*Work speed, smpm*	*Other*
Cylindrical grinding	1650-2000	R 15-21, F 21-30	*Wheel advance*[†] R $\frac{1}{4}$ - $\frac{1}{2}$, F $\frac{1}{12}$ - $\frac{1}{6}$
Surface grinding	1650-2000	12-30*	*Cross feed, mm R 1.6-3.2, F 0.4-0.8*
Tool post grinding			*Traverse, mm/rev*
Cu, bronze, carbon steels	1500-1800	9-11	R 0.8, F 3.0
Ni alloys	1500-1800	9-11	R 0.15, F 0.4
Martensitic stainless steel	1500-1800	33-38	R 0.15, F 0.4
Austenitic stainless steel	1500-1800	9-11	R 0.8, F 1.6

*Table not work speed
[†] Fraction of wheel width per revolution of work

The coating should always be sealed to ensure a better and cleaner initial ground finish. The sealer should be applied before grinding to prevent debris from the grinding operation entering the pores of the coating, and the smoother surface makes it much easier to wash clean afterwards. Removal of all the debris may be a critical factor in the life of a bearing used with a metal sprayed shaft.

OTHER MACHINING METHODS

Planing and shaping
Sprayed coatings on flat surfaces should be finished on a shaper or planer using tools of similar geometry to those employed for turning. Speeds and

feeds should be related to turning practice. It is necessary to remove raised sections or overlapping flash by grinding, using light cuts, before machining commences.

Milling
A milling cutter rotating in a direction opposite to the traverse or work feed is cutting in an upward direction and care is needed to avoid lifting the coating from the substrate. The problem is aggravated in milling keyways and splines. Milling can be performed on flat surfaces provided raised or uneven areas are removed first and that very light cuts are taken. Cutting speeds generally are in the range 20-30 m/min but can be greatly increased for some materials, such as copper alloys. Feed advance is generally 0.2-0.5 mm/rev for roughing and should be reduced according to the finish required.

Linishing
Linishing is often applied where smooth surfaces are desired but tolerances are not critical.

Polishing
Normal equipment and procedures are used. It must be appreciated that the treatment is being carried out on a metallic veneer only mechanically bonded to the substrate. The softer the metal the better the polish so tin, zinc, Babbitt and aluminium are easy to polish, the harder 'red and yellow' alloys are a little more difficult, while the even harder steels, Monel and nickel alloys present great difficulty.

PLASMA SPRAYED COATINGS
Where these are equivalent to the materials considered above machining is undertaken using similar parameters, but other metals and alloys may be sprayed with plasma at normal or reduced pressures. These materials, generally, are finished by grinding rather than turning.

Dry grinding
This is used on many plasma sprayed materials and is quite practical provided that robust equipment is employed with carefully dressed wheels. A grinder which is too light for the work, or a badly dressed wheel causing vibration, will damage the coating and produce a poor finish. Silicon carbide abrasive is normally used for metallic coatings but diamond wheels can be used with advantage on ceramics, cermets and coatings containing tungsten carbides.

Wet grinding
When suitable equipment is available this is preferred to dry grinding. In these circumstances no special difficulties should be encountered as compared with grinding the same materials in other forms, however, the coating has a different structure so variations in wheel type and operating parameters are required. The coatings tend to be denser than those produced by flame spraying so, generally, grinding wheels should be denser and the grain size finer than is used for the latter.

Quality assurance in surfacing

CHAPTER 8

The control of quality in the manufacture of engineering components has always been important but, in recent years, this importance has become more widely recognised. There are several reasons for this emphasis on quality: components must often withstand more exacting service conditions, they must be produced to more demanding economic standards, and they are affected by environmental and legislative factors.

The control of quality of a surface coating is no different in principle from the control of other engineering products but, in practice, coatings require specialised techniques. It is the aim of this chapter to consider these and to outline the main factors that should be taken into account to ensure quality of surfaced components. Examples of application of the principles are given, drawn from weld surfacing and thermal spraying, which are the most widely applied coating processes for the reduction of wear and corrosion. The quality of coatings applied by electrodeposition and by PVD and CVD is covered in the chapters dealing with these processes.

The quality plan

The deposition of a surface coating involves a series of interrelated stages. To achieve good quality repetitively it is essential for each and every stage of this processing cycle to be controlled, with standards clearly defined and rigorously followed. Surface coatings, by their nature, pose problems in testing and, consequently, good quality must be built into the deposition process, rather than attempting to test out poor quality after fabrication has been completed. This emphasis on good housekeeping throughout the processing cycle is the keynote to quality assurance in surfaced parts.

The sequence of operations needed for any particular job should be clearly laid down at the outset to form the 'quality plan': a specification should be placed upon each operation in the process, so that it may be carried out effectively and reproducibly.

Table 8.1 The quality plan

Design	*Can the system be redesigned to reduce the severity of environmental effects?*
	Is the part design suitable for the proposed surfacing technique, or can it be modified to be suitable?
Component	*Is the composition/structure suitable for the proposed surfacing operation?*
	Can the part withstand the necessary preparation?
	Is the component surface free from defects
Surfacing choice	*What is the most suitable coating and deposition process, taking into account:*
	(a) service requirements; *(b) economic criteria;* *(c) equipment and expertise available;* *(d) location of job, etc*
Surfacing procedure	*Process parameters needed*
	Control needed (manual, automated, operator skills)
	Pretreatment required
	Post-treatment required
	Finishing required
Quality assurance	*Quality required for application*
	Defects that may be encountered in chosen surfacing process
	Agreed tests to be used and acceptance standards
	Tolerance to defects and permitted rectification

The above list is typical of the factors that should be considered in setting up the quality plan for a surfacing job; the list is not comprehensive and may contain more (or less) items, depending upon the application.

The elements of a typical quality plan are listed in Table 8.1 and the individual items are discussed below. The preparation of the plan should involve collaboration between all interested parties: ideally, designer, processor and user should be consulted; only if this is done can optimum results be expected. The plan should be appropriate to the component and its service requirements. It need not — in fact it generally will not — aim at perfection, which is rarely obtained and seldom justified. Rather, it should take account of the types of defect that are likely to occur in processing and should evaluate the extent to which such defects can be tolerated for the application.

DESIGN

There is a widespread belief that, if a part is to be surfaced, any design will suffice. This idea is completely wrong. There are certain design principles for parts to be coated and, if these are followed, many potential problems can be avoided. Conversely, incorrect design adds to the difficulties of coating a component and, in the extreme, may preclude it entirely. Thus, incorrect blank dimensions can affect the finished thickness of the deposit and its uniformity or concentricity; these in turn can affect wear resistance and thermal stability in use. Sharp corners and recesses can lead to unsound deposits and, in weld deposition, to excessive local dilution of deposit material by the substrate.

These comments assume that design changes can be made. Although this should be so with original equipment, it may not be feasible to implement modification when surfacing is to be used for repair or reclamation; a compromise on design features is then inevitable. However, even minor modifications, e.g. chamfering sharp corners, should be considered, as they can have a disproportionate benefit in subsequent processing.

COMPONENT/SUBSTRATE REQUIREMENTS

The surface to be coated must be suitable for the proposed coating and deposition method and it must be appropriately treated before surfacing begins.

The composition of the component, and its metallurgical condition, have a direct and important bearing on the success of the finished job. The composition, especially with weld deposition, must be chemically compatible with the proposed deposit material. It is desirable, therefore, to know the composition of the component, either by a suppliers' certificate of analysis or an in-house analysis. This requirement is perhaps rather less important with coatings which do not involve substrate fusion but it may be important to check the metallurgical condition of the substrate to ensure that it is suitable, e.g. if it is too hard it may not be possible to achieve effective roughening during grit blasting in preparation for thermal spraying. Structural condition can be checked by simple metallographic examination supported by hardness tests.

High quality deposition can be obtained only if the component surface is free from defects: porosity, cracks, or inclusions (in or near the surface layers) lead to poor quality coatings. Soundness must therefore be checked carefully, possibly by dye penetrant or magnetic particle testing, or by radiography or ultrasonic inspection.

The substrate must be able to withstand any preparation for coating, as well as the conditions during deposition. Thus, where a coating system requires a clean, rough surface it will involve removal of paint and loose rust by pickling, scraping or wire brushing, vapour degreasing, and then grit blasting the components. To minimise blasting time and ensure adequate adhesion, grit is carried at high gas pressures (~ 0.6 MPa) and the part must not distort under these pressures. For delicate components lesser pressures may have to be used, but this can militate against part throughput and coating adhesion.

The ability of the component to resist reasonable heat input may also be necessary, particularly if surfacing by one of the 'hot' processes. Heat resistance may, in any event, be desirable, as heating of the substrate is often practised in 'cold' deposition processes to improve adhesion. Finally, large differences in coefficient of thermal expansion between component and surfacing material should, if possible, be avoided where problems from contractional stresses on cooling may arise.

Again, the above criteria may be reasonably easy to achieve in the coating of original equipment, but may pose problems in reclamation work where a compromise may be inevitable.

CONSUMABLES FOR WELDING AND THERMAL SPRAYING

The material to be applied may be in the form of rod, wire, cord, or powder, depending upon the process selected. In general, appropriate standards should be placed on composition and physical characteristics, e.g. rod diameter, powder particle size. It is necessary to ensure that any description used is adequate and unique to the material required: makers' brand names and trade names may cover a range of somewhat different alloys. Most suppliers will supply a guarantee of conformity with (or a certificate of) analysis and this should be obtained; failing this the feed materials should be checked in house. In any event, the material must be clearly marked before it is issued for use on the shopfloor.

Where the coating material is presented as a powder it is necessary to make sure that it will feed properly; this entails detailed evaluation of the powder. Chemical analysis, powder shape, size and distribution, apparent density, and flow rate are measured and structure and cleanness may also be assessed. Powder tests such as those listed in Table 8.2 are suitable.

Table 8.2 Methods for evaluation of thermal spraying powders

Property	Method
Powder sampling	Dividing and quartering powder
Apparent powder density	Weight of powder to fill Hall flowmeter
Powder flow rate	Time for powder to flow through Hall flowmeter orifice
Particle size analysis:	
Sieve analysis;	Mechanical sieving (sieves to BS 410)
Subsieve analysis	Coulter counter technique. Sedimentation, elutriation, etc
Particle shape	Low powered microscopy
Powder surface area:	
BET measurement;	Nitrogen adsorption
Others	Air permeability. Calculation for size distribution
Chemical analysis	Standard tests
	Hyrogen loss

These feed materials must be kept in suitable conditions while awaiting use. In particular, care must be taken that they are kept dry; this is commonly achieved by storing in a heated cupboard.

Strict specifications must be placed on the gases that sustain the heating process and transport the coating material. This is necessary not only for good deposit quality but is often essential to ensure low maintenance costs for the equipment.

Grit blasting the surface before thermal spraying requires use of clean, fresh, and sharp grit, with a minimum of fines. It must have sufficient strength to prevent premature fracture, so impairing the effectiveness of the blasting operation. Equally, it should be low in impurities that can be transferred to the component surface and interfere with the spraying operation. The air used must be clean, dry, and oil free; this is best achieved by an oil free air compressor. After grit blasting, the prepared surface should be 'blown off' with dry, oil free air and should be sprayed as soon as practicable. The maximum length of time that can be tolerated before spraying varies according to circumstances but, in normal workshop practice, about four hours would be considered the maximum delay.

SURFACING PRACTICE

Successful surfacing depends on availability of suitable equipment, specification of appropriate operating parameters, and the commitment of the operator.

It is necessary to use reliable equipment, fitted with an adequate range of accurate controls and instrumentation and conforming to proper safety standards. Without these basic criteria there can be little concern with component quality. One aspect of the growing use of automatic surfacing methods is not only that they permit greater consistency in deposition but that they also tend to give greater operator safety. Any equipment must, of course, be properly maintained if it is to function correctly.

The operating parameters for the actual coating operation must be carefully specified, to ensure efficient deposition, acceptable throughput and quality of coating. However, the processing conditions which give the highest efficiency or greatest throughput may not give the desired coating characteristics (or may not be acceptable in the substrate). A degree of compromise may thus be necessary in the choice of operating parameters for a particular job.

Heating coils or cooling jets may be used to control the thermal flux to the component. The specification may call for preheating before deposition, control of the interpass temperature, or heating after deposition. With all these operations suitable pyrometric equipment should be made available to monitor and record the temperatures.

Acceptable and uniform results from manual deposition are dependent upon operator skill. Proper training in the process is therefore essential, and there should be periodic qualification tests to ensure that skills are maintained.

In summary, it may be said that good practice in the actual process of deposition is the key to production of satisfactory coatings. A reliable and resourceful operator can eliminate many hazards in the process but there are additional requirements if reproducible quality is to be obtained: suitable equipment, specification (and following) of correct process parameters and maintenance of accurate and detailed records.

FINISHING

The finishing operation can play a vital role in the quality of a coating (and on the economics of production).

Table 8.3 Typical checklist for processing parameters (weld deposition)

Component	Identification/description
Work required	Brief description Finishing requirements
Substrate	Material Specification Condition Preparation required
Preheat treatment	Preheat temperature, °C, minimum and maximum Interpass temperature, °C, minimum and maximum
Description	Process Consumables: rod/wire size, mm Thickness: mm : number of runs Welding parameters: Electrode voltage, DC Electrode polarity Current, A Voltage, V Speed, mm/sec Electrode angle Wearing parameters
Postheat treatment	Heat treatment temperature, °C Heating rate, °C/hr Soaking time, hr Cooling rate, °C/hr
Inspection and testing	Purpose Methods/tests Acceptance level
Special instructions	Detail (e.g. operator qualifications)

Notes: The above list is not intended to be comprehensive but simply to illustrate the type of items to be included. A similar checklist, suitably modified, would be used for thermal spraying operations

Heat treatment and stress relieving operations may be necessary and these must be accurately specified, controlled and recorded. Surface machining, grinding or polishing of the deposit may be necessary to meet tolerances or produce smooth finishes; if the deposit is very hard, diamond grinding may be required. Care must be taken to lay down specifications on the finishing consumables which will ensure that the grades of material, *e.g.* diamond or lapping compounds, are suitable for the operation.

Should it be necessary to seal porosity in a thermally sprayed coating, this would normally be carried out with varnish, epoxy resin, or metal infiltration. The actual method of conducting the sealing can be vital in its success which, in turn, calls for use of a clearly defined and properly controlled procedure.

CHECKLIST OF OPERATIONS
To ensure that all necessary information and instructions are recorded it is desirable to use a checklist to summarise requirements for any surfacing job, see Table 8.3.

Testing and inspection of coated components

The aim of testing surfaced coatings should be to ensure that the operation has been correctly carried out, and not to 'test out' poor components. This distinction may seem trivial but it implies a fundamental difference in approach.

There are many criteria by which coatings can be judged, Table 8.4, and many techniques are available to assess them, Table 8.5. Some of these can be applied without damaging the component but others cannot; furthermore, responses to many of the techniques will be affected by the substrate.

Table 8.4 Potential criteria for engineering coatings

Adhesion to substrate	*Electrical properties*	*Structure*
Chemical composition	*Evenness*	*Surface finish*
Chemical reactivity	*Frictional properties*	*Thermal properties*
Cohesive strength	*Hardness*	*Thickness*
Covering power	*Machinability*	*Wear resistance*
Edge definition	*Porosity*	

Table 8.5 Typical tests applicable to coatings

Generally non-destructive	*Generally destructive*
Visual examination	Bend testing
Dimensional checking	Adhesion testing
Dye penetrant inspection	Peel testing
Magnetic crack detection	Proof machining
Eddy current testing	Chisel testing
Acoustic methods	Chemical analysis
Radiography	Metallography
Hardness testing	Porosity measurement
Ultrasonic testing	

For any particular application only some of the criteria will be relevant. What is essential is that those involved — essentially the coater and the customer — agree on what criteria are to be used to judge any particular coating. To avoid needless ambiguity, this agreement should be reached before deposition starts.

The result of testing procedure can vary according to the skill of the operator and the techniques used. It is therefore desirable not only to define the test conditions precisely but also to audit the procedure (and the operator) regularly.

Different techniques produce deposits with different characteristics; this calls for differences in the detailed test procedure. This is illustrated below by describing testing of weld deposited and sprayed coatings. These represent two distinct types of deposit, the former are fully dense and integral with the substrate but the latter usually contain some porosity and tend not to be metallurgically bonded to the substrate.

Weld deposited coatings

If deposition is manually controlled the welder will be responsible for initial checks that the correct amount of deposit has been applied to the correct place; this requires simple measuring gauges. That these are simple instruments highlights one general feature of good inspection: the tool or the instrument selected should be appropriate to both test and coating.

The inspector normally examines a welded deposit in the following sequence:

1. Visual examination for freedom from obvious defects, such as cracks and porosity;
2. Dye penetrant inspection as an aid to identification of cracks; care is needed in interpretation of the result;
3. Dimensional checks for size, distortion, *etc.*
4. Hardness and non-destructive testing according to the application. Use of radiographic examination is not widespread and tends to be used only for certain specialised applications, *e.g.* surfacing of some nuclear components.

As incorrect grinding can cause surface cracking of hard deposits, a final dimensional check and a dye penetrant test are useful, after finish machining and grinding.

In seeking to set the appropriate quality standard it is necessary to evaluate the more commonly occurring defects in weld deposition, Table 8.6. Excluding dimensional errors these are discontinuities and are of three main types: cracks, porosity, and inclusions.

Cracks may be of various kinds. Contraction cracks arise as a consequence of the low ductility, and possibly low tensile strength, of many hardfacing alloys, which make them unable to withstand the stresses upon cooling of

Table 8.6 Possible defects in weld surfaced and thermally sprayed coatings

Defect	Characteristic	Possible causes
Weld deposited coatings		
Cracks	Contraction	Low ductility or low tensile strength in coating material
	Fine surface cracks	Unsuitable grinding wheel or insufficient cooling during grinding
	Lifting at interface	Poor welding technique with failure to disrupt oxide film
	Lifting within substrate	Embrittlement of substrate during welding
Porosity	Associated with final molten pool	Faulty crater filling
	Randomly dispersed	Flux contamination, faulty flame adjustment (gas welding), damp electrode coverings (arc welding), etc
Inclusions		Incorrect welding conditions or use of dirty substrate material
Thermally sprayed coatings		
Structural defects	Excess porosity	Poor spraying practice
	Oxide inclusions	Pick-up during spraying or excessive surface oxide on feed material
	Unmelted particles	Incorrect powder quality or spraying conditions
	Sooty deposits	Insufficient interval between melting and boiling points in coating material
Poor adhesion		Poor practice and particularly poor substrate preparation

the surfaced component. Such cracking may be made worse by stress raising defects in the coating and by post-weld straightening to rectify any distortion. The presence of contraction cracks may be accepted as a necessary concomitant to the use of the most wear resistant deposit material and, for certain applications, they may not impair service performance.

Their presence rarely involves a risk of the deposit breaking away from the substrate, provided there is no hardening of the heat affected zone and the welding operation has given a satisfactory bond to the substrate. However, such cracks would not be tolerated in the sealing surfaces of valves and sealing rings, in surfaces subject to erosion, and in surfaces that must give corrosion, as well as wear, protection to the substrate.

Fine surface cracking or checking can occur during grinding of hard deposits if grinding conditions are unsuitable or there is insufficient coolant. Such grinding cracks are usually shallow but may propagate with thermal or mechanical stress; they are therefore subject to the same restrictions as contraction cracks.

The deposit lifting from the substrate constitutes another type of cracking. Cracks may occur at the interface between the deposit and the substrate and may stem from a poor welding technique resulting in a failure to develop a good bond, or to failure to disrupt surface oxide films before welding begins. Alternatively, cracks may occur in the substrate, just below the interface with the coating; these are a result of embrittlement of the substrate during welding and can be avoided by ensuring that a correct thermal treatment is given. As the deposit may detach itself during service, such 'lifting' defects cannot be accepted for any application.

Porosity can have various causes. If associated with the final molten pool it may be a result of faulty crater filling. When randomly dispersed it can be a result of, for example, flux contamination, faulty flame adjustment (in gas welding), or damp electrode coverings (in arc welding). It can also be caused by chemical reaction in the molten pool. Porosity may range from a few specks to visible spongy metal, so that a form of graded specification should be used; the degree of acceptance can then be related to the application. It may also be desirable to vary the acceptance in one part of the structure than in another. Any such variations should be shown in the acceptance standards.

For some applications visual examination for porosity may be sufficient; for more critical service dye penetrant examination is used. It may be unnecessary, and probably uneconomic, to specify a 'zero bleed' acceptance standard; moreover, grinding of certain hardfacing alloys may tend to pluck out hard, needle-shaped carbides which dye penetrant testing will pick up. For these reasons the standard for and interpretation of dye penetrant testing must be appropriately set.

Inclusions can range from a size which is obvious in general examination to that which can be detected only by dye penetrant testing. They may be caused by use of dirty substrate material but more usually they arise from incorrect welding parameters or technique. Their identification and acceptance standard can conveniently be grouped with those criteria for porosity.

Where defects are found the designer will decide whether the deposit can be rectified. If weld deposition has originally been used it would be normal to use welding again. The following points should be taken into account when making such repairs:

1. Repairs should be made at the proof grinding stage where there is usually still sufficient machining allowance to deal with any further distortion or oxidation;
2. The same metallurgical conditions should generally be specified as for the original deposit;
3. The defective area must be removed completely, down to sound metal, verified by dye penetrant testing. A shallow rounded depression should be produced and not one that is narrow and deep or has sharp corners. If the substrate metal has to be replaced, the profile should be restored by a compatible electrode (and not by hardfacing alloy) before recoating.

THERMALLY SPRAYED COATINGS

The dimensions of the coated product must be correctly held to the customer's specification and these — as with welded deposits — must be checked. Again, a micrometer is usually sufficient for the accuracy demanded, although electronic displacement instruments (based upon eddy current or magnetic techniques) are coming into use.

The quality of the coating is vital. As coatings become less dense and less homogeneous, and as adhesion to the substrate tends towards mechanical rather than metallurgical, existing NDT techniques become less reliable; greater emphasis must consequently be given to destructive techniques such as adhesion tests and metallography. Destructive testing on a small percentage of components is acceptable where large numbers of similar articles are being processed. However, many thermal spraying jobs involve coating of small numbers of components: often, in reclamation work, one part might be processed. In these circumstances it is necessary to coat a test standard or coupon. The difficulty is then to ensure that the coating on the components is similar. This requires not only consistency in spraying conditions but also an appreciation of the effects of, for example, differences in component section thickness and cooling rate on deposit structure and properties.

Metallographic examination is widely used to assess deposit quality, but care must be taken in preparation of the section and in interpretation of the results to avoid misleading conclusions. At its best, such examination can show up defects, give an indication of porosity and highlight features of the interface.

Hardness, being easily determined, is commonly measured on coatings but its usefulness is limited. Tests of the Brinell or Vickers type give a general indication of coating quality, but the presence of porosity will lead to an apparently low hardness compared with that of similar solid material. Additionally, the presence of a hard substrate which supports a thin, soft surface film will give a hardness that is affected by deposit thickness. These aspects cast doubt upon the absolute use of hardness measurements but may be useful for comparative purposes. Microhardness may also be used as a comparative check that the correct deposition conditions have been achieved.

The importance of coating adhesion and cohesion is such that a large number of methods of measurement have been tried. Many, such as chisel and scratch tests, are of a qualitative or semi-quantitative nature and are used with a skilled operator only for comparison of coatings. Others, however, set out to be quantitative and many testpiece designs and pulling arrangements have been tried. Figure 8.1 shows a typical example of testpiece design. To assess cohesion within the coating, two accurately machined tubes, which have been sprayed on their outer surfaces, may be pulled apart. To determine adhesion two cylinders are used, one being coated and resin-bonded to the other; these are then pulled apart. There are, however, severe limitations in tests of this type. If adhesives are used the result may be affected by the type of adhesive, the degree of penetration into a porous coating, and the conditions used during curing. The tests may, in any event, be affected by alignment of the specimen in the test machine and by strain rate variables between tests. Such tests must thus be interpreted with caution; at best they represent a degree of consistency, so that gross variations in readings may indicate poor production conditions. For these reasons there has been continuing interest in the development of a reliable non-destructive method to measure coating adhesion. Despite much effort, no fully satisfactory method has yet been proved but recent work, based upon use of an ultrasonic method, gives promise of being developed into a useful technique.

The NDT methods to assess general coating quality have had only limited success with sprayed deposits. Surface finish measurements may be useful to indicate batch to batch differences in coatings: it has been suggested that surface roughness can be developed as a quick and reliable method of checking coating quality. Crack detection by magnetic or dye penetrant techniques often fails to detect isolated defects against the general presence of porosity. Eddy current and ultrasonic inspection are little used for sprayed coatings, which tend to be heterogeneous and give a high noise-to-signal ratio. Radiography can be used but with some uncertainty in interpretation.

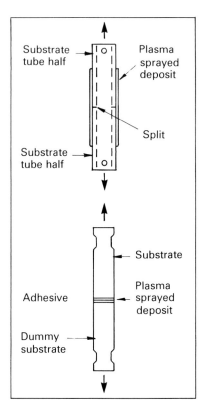

8.1 *Test arrangements for sprayed coatings:* a) *Cohesion;* b) *Adhesion.*

Table 8.6 includes defects which may occur in sprayed coatings. Metallography will highlight many of the defects which can arise from bad spraying procedures: excessive porosity, high oxide content, the presence of unmelted particles, and possibly poor bonding of coating to substrate. Most sprayed deposits have pores arising from shrinkage and degassing after deposition; the former gives acicular shaped pores, the latter more rounded ones. However, poor spraying practice can lead to a condition of gross porosity which must be avoided. Oxide inclusions can arise from pick-up during spraying — especially in the oxygas or arc methods — or from excessive surface oxide on a (powder) feed material. The latter may be controlled by attention to quality and particle size during powder manufacture. Where the oxide content from pick-up is unacceptably high it may be reduced by use of a shroud to protect the workpiece; this is quite common. Alternatively, deposition may be carried out in an inert gas chamber. The presence of unmelted particles indicates incorrect powder quality and/or spraying conditions. One of the advantages of using a solid, e.g. wire, feed is that the coating material must be melted before it is projected at the substrate. There is a need for a reasonable temperature interval between the melting and boiling points of the coating material, otherwise excessive evaporation can occur which will lead to a 'sooty' type of loosely bonded deposit. It should be emphasised that these defects will not occur with good spraying practice and, indeed, usually cause few problems.

Inadequate adhesion of coating to substrate is a defect which cannot be tolerated. There are certain deposits which inherently give problems with bonding to certain substrates and an intermediate (or bond) coat may be used to overcome problems. Equally, certain part spraying treatments can be used to enhance the integrity of the deposit. In good practice, however, poor adhesion most commonly stems from poor substrate (component) preparation, and even slight departure from the recommended practice — especially in grit blasting — can give poor results. If such problems exist the defective coating must be stripped and the part recoated; at the same time the operating practice should be critically reviewed. If a coating is subjected to a finishing operation, its ability to withstand this is a good empirical test of deposit adhesion.

Conclusions

The aim of this chapter has been to present the view that quality assurance of surface components depends not upon sophisticated and expensive test methods, but upon simple, but strict, control of each stage of the processing cycle. Starting at design, each stage should be carefully considered, defined, and carried out as specified.

Faulty materials or design, or incorrect surfacing practice, can produce a poor deposit which may be difficult to detect. The philosophy must therefore be that poor quality cannot be tested out; good quality must be built in so that inspection then becomes a verification that good practice has been observed.

Safe working in surfacing

CHAPTER 9

The Health & Safety at Work Etc Act 1974 put the duty on employers, designers and controllers to provide premises and plant which are safe and without risk to health when properly used. It also charges employees to take reasonable care for their own health and safety and for that of others, and to co-operate with employers as far as necessary to enable duties or requirements to be performed and complied with.

This chapter briefly describes hazards encountered in surfacing, and the steps which should be taken to overcome them and achieve safe operation. It is divided into three sections: surfacing by welding and thermal spraying; vacuum deposition; electrodeposition.

The bibliography at the end of this chapter lists the principal UK standards and legislation. Other countries have a similar range of documentation. A more detailed treatment is given in The Welding Institute's book 'Health and safety in welding and allied processes': the third edition was published in 1983 and so changes since then are not covered. Most of these are of little relevance to surfacing, except that Threshold Limit Values have been superseded by Occupational Exposure Limits.

Surfacing by welding and thermal spraying
COMPRESSED GASES

Storage
Cylinders containing compressed gases, Table 9.1, must always be handled with care and stored in well ventilated areas away from the working area whenever possible. Oxygen and fuel gas cylinders should be separated, and no smoking allowed in or near stores. Fire brigades should be kept informed of the location of cylinder stores.

Table 9.1 Cylinder and equipment identification in the United Kingdom

	Thread, male/female	Colour
Cylinder		
Air	RH F	Grey
Argon	RH F	Blue
Argon + oxygen	RH F	Blue, black band
Argon + CO_2	RH F	Blue, green band
Helium	RH F (usually US thread size)	Light brown
Nitrogen	RH F	Grey, black collar
Oxygen	RH F	Black
Hydrogen	LH F	Red
Propane	LH F	Red
Acetylene	LH F	Maroon
Carbon dioxide	RH M	Black
Equipment		
Fuel gas coupling	LH	Red, except propane hose — orange
Other gas coupling	RH	Inert gas — black Oxygen — black

Acetylene cylinders
Suppliers attach a label with the following wording to each cylinder:

> 'If an acetylene cylinder is seriously heated accidentally, or becomes hot because of flashback: shut the valve, detach regulator, remove cylinder outdoors at once, spray with water to cool, keep cool with water. Leave cylinder outdoors, advise suppliers immediately.'

This warning is because acetylene under pressure can undergo a chemical reaction which releases heat and tends to increase the pressure. This reaction can be triggered off by a flame entering the cylinder from a flashback, or by general heating of the cylinder, for example in a fire.

Cylinders are constructed to a specification which ensures that such a reaction should be self-limiting: the reaction could continue, for example, as a result of continued heating in a fire, or if there should be a defect in the construction or maintenance of a cylinder. The reaction can be arrested by cooling the cylinder, hence the advice above. Safety devices (fusible plugs and/or bursting discs) may be provided to limit pressure increase; when these operate there will be a massive release of highly flammable gas. If they fail, pressure may increase to the point at which the cylinder bursts.

Equipment

Regulators and other equipment must be suitable for the gas, with undamaged connectors, and a working pressure rating equal to or exceeding the maximum cylinder filling pressure.

Flame equipment should always be fitted with a flashback arrestor.

Gas hoses and their connectors must be inspected regularly to ensure that they remain in good condition; particular attention must be paid to trailing hoses. If repairs are necessary, they must be made with the correct connectors and hose clips.

Where portable welding equipment is in use, cylinders must be carried on a properly designed trolley.

All equipment, particularly that used with oxygen, must be kept clean and free from oil and grease.

If working in a confined space, provide ventilation as needed to avoid accumulation of excess concentrations of burnt fuel gas or shielding gas.

Check equipment for leaks after assembly.

Always:

- Check that the right gas is being used to suit the equipment and job.
- Avoid smoking when handling or connecting fuel gas or oxygen cylinders.
- Secure cylinders so that they cannot fall over.
- Store and use acetylene and propane cylinders upright; use carbon dioxide cylinders upright.
- Protect cylinders from heating, impact, or accidental contact with arcs.
- Keep connections clean and dry.
- Use correct regulators and fittings.
- Avoid use of cylinders as work supports or as rollers.
- Examine equipment for damage or wear before assembly.
- After assembly of equipment, check for leaks (with a 0.5% detergent solution for flame equipment).

FIRE

Flames and arcs

Use of flames or arcs in welding, preheating, *etc*, presents an obvious fire hazard and appropriate precautions should be observed. No combustible material should be allowed to be used or stored near the working area and workbenches; partitions and other fixtures should be made in fireproof materials. The possible hazards from weld spatter and from heated overspray particles should not be overlooked.

Metal powders

Finely divided metal powders may constitute a fire or explosion hazard if allowed to collect in sufficient quantity. The work space should be designed to reduce dust accumulations and should be cleaned regularly. Ducting should be designed so that it can be cleaned when necessary.

All motors, switchgear, fans, and other electrical equipment should be properly earthed and should preferably be of the flameproof type.

Always:
- Avoid excessive release of fuel, gas or oxygen into the atmosphere.
- Remove flammable materials from the working area.
- Where adjacent flammable material cannot be removed, station an observer to deal with any outbreak of fire.
- Check for fire about an hour after work has finished.
- Check emergency escape route.
- Use a tray to catch overspray where possible, preferably sand or water filled.
- Ensure that appropriate firefighting equipment is at hand and that all personnel know how to use it.

Vessels with residual flammable material

A used or contaminated vessel, tank, drum, pipe, or other closed installation which has at any time contained a flammable liquid or dust is liable to explode if a surfacing process is operated in or on it. The risk can be removed only by thorough cleaning or filling with an inert gas.

Always:
- Obtain a certificate from a competent person that any used or contaminated vessel is free from risk of explosion before preparing to work on it.
- Where a permit to work system is in operation, ensure that all instructions are understood and obeyed.

FUMES AND DUST

With all welding operations, fume is produced; with spraying there will be both fume (particle diameter $<0.5\mu m$) and dust particles (diameter $\geqslant 0.5\mu m$).

Fumes and dust can arise from:

1 Airborne particles of the spray material;
2 Chemical compounds of the surfacing material, and of the parent metal, such as oxides;
3 Gaseous products of decomposition of fluxes and electrode coverings;
4 Products of combustion of the gases used, *i.e.* carbon monoxide, carbon dioxide, oxides of nitrogen, water vapour, *etc.*

Occupational Exposure Limits

The maximum concentrations to which workers may be exposed for most of the dust and fumes which are met in surfacing operations have been laid down in published Occupational Exposure Limits (OEL) figures: these have replaced the former Threshold Limit Values (TLV). Some materials are more harmful than others: for instance, cadmium, chromium, cobalt, copper, manganese, nickel, and zinc (this is shown by a lower OEL; obtain and observe any special recommendations of the consumable manufacturers). Special regulations apply where lead is present.

Where any doubt exists as to the danger of a particular material, or of the effectiveness of ventilation and extraction systems, the concentration should be measured. Even if levels are below the OELs this may not meet the requirement in the UK of the Health and Safety Executive that pollutants should be reduced to the lowest level which is reasonably practicable.

Ventilation

Adequate ventilation of the working area is necessary to remove fumes; the position of ventilation inlets and outlets should be arranged so that the fumes are taken away from the operator. The exhaust must be placed to avoid recirculation of fumes to work areas.

Dust extraction requires relatively high air velocities and the use of an extraction outlet as near as possible to the workpiece will give the highest efficiency. In addition to extraction it may be necessary to provide for dust collection, especially with materials having a low OEL. Use of water-washed spray booths is now common practice for this purpose but other methods are also available and each case must be judged individually.

Personal dust and fume protection
This can be used to supplement ventilation to give adequate reduction of dust and fume; the options available are:

1   Dust respirator with replaceable filter;
2   Disposable dust respirator;
3   Air fed helmet;
4   Helmet with fan and filter (positive pressure powered dust respirator).

Always:
- Keep degreasing fluids and vapour away from the work area.
- Check for possible toxic hazards from parent metal or consumables.
- Where necessary use a local fume extractor, repositioned as work proceeds.
- Ventilate confined spaces with extractor fans exhausted to open air.
- Supplement fume extraction with personal protection where necessary.

Work in enclosed spaces
If surfacing is to be carried out in an enclosed space, special care is needed because of the risk of a build-up of fume or gas.

Always:
- Ensure adequate ventilation, with personal fume protection where necessary.
- Ensure that any used vessel, tank, drum, or pipe does not contain flammable, poisonous, or corrosive material (see 'Vessels with residual flammable material', above).
- Ensure that gas cylinders are not taken into an enclosed space.
- Check equipment for gas leaks with special care.
- Ensure that any emergency which may arise will be noticed.
- Check by rehearsal that the worker can be rescued should an emergency arise.
- At the end of work periods, shut off gas supply valves and withdraw hoses and equipment.

NOISE
Plasma spraying and, to a lesser but still significant extent, wire spraying operations give rise to intense high frequency noise which can cause damage to hearing. Operators must wear ear protection: either re-usable or disposable ear plugs or ear muffs. Soundproof screening may be necessary in some areas for protection of adjacent personnel.

Always:
- Wear suitable ear protection when exposed to loud noise.

RADIATION
All surfacing processes which make use of a flame or an electric arc emit radiation in the form of heat, visible light, and ultraviolet radiation.

Radiation to the eye
The ultraviolet radiation from an arc, and to a lesser extent from a flame, causes a painful irritation of the outer surface of the eyeball known as arc eye. Exposure for only a few seconds to the light from a high current welding arc can produce severe discomfort, starting a few hours after the exposure; fortunately, most cases recover completely in a day or two with no permanent ill effects.

The visible light is usually too intense to allow comfortable viewing. The potential hazard here is to the retina, the light sensitive surface at the rear of the eye, which could be permanently damaged by viewing a bright source such as an arc: blue light is the most harmful. The infrared radiation may also be intense enough to be harmful in rare circumstances.

Protection of the operator is afforded by a suitable filter which cuts out ultraviolet and infrared, and reduces visible light to a comfortable level. A suitable specification is BS 679, and see also ISO 4850 which identifies filters by a shade number; the higher the number the darker the filter. Table 9.2 gives recommended shades. Filters to this specification are approved under the UK Protection of Eyes Regulations, which require the provision of protection in processes using an exposed electric arc, *etc*. It is customary to protect the filter against spatter damage with clear plastic cover sheets on either side.

Table 9.2 Protective viewing filters for use when surfacing (based on BS 679 recommendations)

Process		Filter
Gas welding, spraying, and surface fusing		5/GW
		6/GW
Gas welding using flux		5/GWF
Gas welding on large areas, continuous working		6/GW
		7/GW
Manual metal arc welding (covered electrodes)	< 100A	8/EW
		9/EW
100-300A		10/EW
		11/EW
	> 300A	12/EW
		13/EW
		14/EW
MIG welding	< 200A	10/EW
		11/EW
	> 200A	12/EW
		13/EW
		14/EW
TIG and plasma arc welding	< 15A	8/EW
	15-75A	9/EW
	75-100A	10/EW
	100-200A	11/EW
	200-250A	12/EW
	250-300A	13/EW
		14/EW
Plasma spraying*		9/EW-15/EW
Arc spraying*		6/EW
		9/GW

*There is considerable variation in equipment from different manufacturers in respect of the luminosity of flame or arc, and the extent to which it is shielded from the operator's view; see the manufacturer's recommendations or, if this is not possible, choose a shade giving a clear comfortable view.

In addition to radiation, a filter also prevents stray particles of coating materials, *etc*, from entering the eye and burning or otherwise injuring its outer surface: there is now a requirement for impact resistance in protective helmets, *etc*.

For light duty gas processes, where face protection (see next section) is not required, the filters may be used in suitable goggles. Instrument panels, dials, *etc*, should be positioned so that they do not reflect light from the arc to the operator or anybody else.

Walls, ceilings, and working surfaces should be painted so as to reduce amounts of ultraviolet radiation. This can be achieved by paint of almost any colour except metallic: but blue should be avoided, to minimise visible light hazards. Pastel shades of matt emulsion paint are effective and easily applied. The polished surfaces of metal or glass, gloss paint finishes, and water in wet booths reflect an image of the arc which is often distracting, and may be harmful, and so should be avoided by carefully positioning the work.

Protection of the eyes of others nearby (required by the UK Protection of Eyes Regulations) may be given by fixed or portable screens, or by personal goggles or face shield. Screens can take a number of forms:

1. Opaque solid (hardboard or sheet metal);
2. Opaque flexible (fireproof heavy canvas);
3. Transparent or translucent coloured flexible plastic, transmission corresponding to viewing filters;
4. Translucent uncoloured flexible or rigid plastic or glass; most such materials transmit only diffused visible light, greatly attenuating infrared and ultraviolet.

Likewise, clear plastic or glass spectacles, goggles, or face shield will protect against most brief or minor exposures to ultraviolet.

Radiation to the face
The intense ultraviolet from the electric arc will rapidly cause 'sunburn' on the unprotected face, so a helmet or hand shield is used, with the appropriate filter.

Radiation to the body
In all electric arc processes ultraviolet is produced which, on prolonged exposure, can cause skin burns; suitable precautions are necessary to guard against both direct and indirect (reflected) exposure.

Particular attention must be paid to the hands and neck and it should be noted that normal clothing does not necessarily give adequate protection. For instance, with the high intensity emission from plasma torches and any process where high currents are used it is possible to experience ultraviolet burns through a nylon or thin cotton shirt. Hence, proper overalls which fasten up to the neck, gloves, and face masks (see above) should be worn. Fabrics should be close-woven cotton, which will not readily catch fire. White overalls should not be worn when arc processes are in use because they reflect too much light.

With welding it may be necessary to provide additional operator protection if work is to be carried out on large workpieces with high preheat temperatures.

Laser radiation
Surfacing can be carried out using a laser which supplies enough energy (2-10kW) in the form of a beam of radiation to melt powder to form a coating on the surface of a parent metal. The infrared power beam can burn eye or skin, even after reflection from a dull or diffuse surface. The usual safety procedure is to enclose the whole work zone in material which will not transmit infrared nor be seriously damaged or set on fire by accidental impingement of the beam. Metal is often used, with clear plastic windows (even clear plastic strongly absorbs infrared), or the whole enclosure may be made of clear plastic for demonstration or experimental work. Many equipments use an auxiliary visible light (red) laser beam to mark the beam path for system alignment, which could damage the retina of the eye if it is allowed to enter the pupil for any length of time; care is needed not to look along the beam path during alignment.

The Protection of Eyes Regulations apply in the UK to laser equipment.

Always:
- Use helmet, hand shield, or goggles fitted with a suitable filter and clear protective sheet.
- Check that helmet, hand shield, or goggles do not admit light as a result of damage.
- Wear appropriate protective clothing to prevent burns.
- Warn bystanders before striking an arc or operating a laser whose beam is not enclosed.
- Screen working area to protect other working in the vicinity.

ELECTRIC SHOCK
All electrical equipment must be properly installed by competent personnel and must be regularly inspected and maintained. Particular attention must be given to trailing leads, especially primary (mains) leads, to ensure that their insulation and end connections remain in satisfactory condition.

Equipment casings must be earthed in the normal manner, with the usual mains earth; in addition, the work must be earthed independently by a cable and earth connection which is capable of carrying the full current output of the set (the welding return connection is not an earth lead).

High frequency supplies used to start the arc in TIG and plasma processes can cause deep burns if they are allowed to contact the skin. Care is needed to:

1. Maintain all insulation in good condition;
2. Ensure that the work is earthed by a reasonably short lead;
3. Avoid operation of the high frequency unit without its covers in position, as it uses a potentially lethal high voltage supply.

Always:
- Earth workpiece separately.
- Check for damage to insulation on electrode holders, torches, guns, cables, and connectors.
- Keep working area dry.
- Provide emergency means close to the worker to cut off the current when out of reach of the controls.
- When working where a shock could cause a fall, provide guard rails or other protection.
- Check that equipment casing is correctly earthed on installation.
- Switch off equipment before cleaning or dismantling guns.

SURFACE PREPARATION

Abrasive blasting

Abrasive blasting is the most commonly used method of surface preparation. A major hazard would result from inhalation of any silica in the abrasive particles, so commercial blasting media do not contain free silica (in the form of sand, for example) and its use is not permitted under UK law. Instead, materials such as aluminium oxide, chilled iron or steel grit or shot, bauxilite or glass beads are used in various proprietary formulations.

Although the abrasive is therefore non-hazardous, it will have a considerable nuisance value as well as being detrimental to other machinery in the same workshop. Any blast cabinet installation should be fitted with an adequate dust collector and effective door seals, which should be maintained in good working order. Safety door interlock switches should be fitted on automatic installations, and are desirable on manual equipment.

A recommended maintenance schedule is:

Daily
Check seals and toggle catches
Check media for fill and contamination
Check nozzle and air jet for wear
Check gun hose connectors

Weekly
Check hoses
Check dust collector seal
Check pick-up tube for wear

Monthly
Check electrical cables and connections
Check dust collector rotational speed and other functions
Check visual inspection

Electrostatic charges may build up on gun or workpiece if either is not earthed, but the gun will normally have its own earth lead with the supply hoses, and the work is readily earthed except where it is held in the operator's gloved hand, in which case a lead can be taken from a convenient point, usually at the armhole, to a ring fitting outside the glove.

With work on a larger scale where the operator enters the cabinet, or for open site work, the operator must use gauntlets, overalls, and a suitable helmet providing eye protection with a supply of fresh air (0.17 m^3/min).

Always:
- Use only approved abrasive media.
- Avoid excessive contamination of media.
- Check seals regularly.
- Check dust collectors regularly.
- Ensure adequate fresh air supply and protective clothing for operators working within cabinets or on open site.

Solvent degreasing

Solvents available for industrial degreasing may be hazardous because of flammability and/or toxic vapours. An example of a flammable solvent is toluene, which also has harmful vapours; UK regulations specify the hazard control measures to be adopted. Non-flammable solvents, such as carbon tetrachloride, trichloroethylene, and perchloroethylene, evolve vapours which are themselves toxic, and will also be decomposed by heat or ultraviolet light from an arc, with the evolution of more toxic gases such as phosgene. Stabilised 1.1.1. trichloroethylene is the least toxic of the chlorinated solvents in common industrial use, but is still liable to decomposition. It can also react explosively if left in contact with aluminium for any length of time.

Ventilation should effectively remove fumes; where degreasing must be carried out near surfacing operations, the general ventilation flow should lead away from the surfacing area towards the degreasing bay. The exhaust must be placed to avoid recirculation of fumes to work areas. A solvent tank or vapour degreasing plant is preferred for consistent results and straightforward hazard control; where manual degreasing is used, safe disposal of used swabs is essential.

Always:
- Keep degreasing fluids away from the work area.
- Ensure that degreasing vapour does not reach surfacing operations.
- Ensure safe disposal of used solvents.

Grinding

Preparation by grinding is subject to the Abrasive Wheels Regulations and the Protection of Eyes Regulations; the latter require approved eye protection, such as to BS 2092: 1987. The Abrasive Wheels Regulations require the maximum permissible speeds of wheels to be specified, machines to be marked with their operating speeds, and, where practicable, guards and rests to be fitted. Wheels must be fitted by competent persons, specifically trained and appointed.

Always:
- Use suitable eye protection when deslagging, chipping, or grinding.
- Ensure abrasive wheels are correctly fitted.

Vacuum deposition

PVD TECHNIQUES

All these processes involve vacuum equipment, to which very little hazard is attached. Rotary pumps should not exhaust directly into the work room because there is often a discharge of fine oil droplets from such pumps during the initial pumping down stage. Diffusion pumps pose few dangers other than the risk of a (thermal) burn from the heater. Some vacuum gauges operate at potentials of a few kilovolts, but they are generally well protected by the manufacturer. Liquid nitrogen is sometimes used and the operator should be adequately protected against splashes by wearing gloves and goggles (preferably a face mask) when pouring; there should be ample ventilation to prevent build-up of asphyxiant during pouring and general operation. Care is needed to avoid risk arising from evaporation of nitrogen and condensation of oxygen in open containers leading to an oxygen rich liquid. There will be some special risks arising from reactive gases when reactive techniques are used, but these risks are often specific to the gases involved; the supplier's advice should be sought and closely followed.

There is a risk of electrical burn and shock when radio frequency (RF), or ionising processes (e.g. ion plating, sputtering) are used; in both considerable power is available and fatal injury could result. Prevention of accidents is a matter of design, in particular provision of interlocking to prohibit access to 'live' equipment, and adopting a suitable operating procedure. Modified or home made equipment should be checked by the appropriate authority.

CVD TECHNIQUES
Nearly all CVD techniques involve high temperatures and toxic or carcinogenic gases such as carbonyls or organo-metallic compounds. Although such gases are not directly subject to statutory regulations in the UK, nevertheless, their use must be strictly controlled and advice must be sought from the appropriate authority before embarking on any work in this field. It is suggested that, in the UK, the Health and Safety Executive or suppliers of the materials be approached first.

Electrodeposition
WORKFLOW
The objective of an electroplating installation is to apply a specific finish to a variety of surfaces. These surfaces can be reactive to various chemicals and a thorough knowledge of the initial, intermediate and final surface conditions is required if work is to be accomplished safely and competently. For example, a plating shop might not be involved in cadmium plating and might not have a means of disposing of cadmium containing effluent. If, however, an original surface has been plated, soldered or welded with cadmium containing materials, these materials might enter the effluent cycle or be handled by the workers and special protective measures might have to be considered for them.

In a similar way the normal pickling process based on hydrochloric or sulphuric acid does not present substantial atmospheric pollution problems. If, however, hydrofluoric acid has to be used for cleaning aluminium, the emission of hydrofluoric acid into atmosphere demands maintenance of rigorous standards.

Cyanide containing chemical residues on the surface do not produce atmospheric hazards. If, however, such surfaces are immersed in acid solutions the cyanide radical is reduced to volatile hydrocyanic acid which presents acute inhalation dangers.

It is thus extremely important that the chemical state of the surface and its processing cycle is known. Equally important, however, is the layout of the processing cycle in the plating shop. This can create problems of dangerous cross-contaminations, gas evolution or work spoilage.

ENGINEERING
Although often neglected the condition of engineering appliances in the plating shop can contribute significantly to hazards. Critical areas are construction, transmission and extraction. Problems are often caused or aggravated by the wet and corrosive conditions which prevail in the plating shop and for these reasons its construction and the equipment in it are important. It must also be remembered that emission to atmosphere or environment through an opening in the roof or leaking draining system constitutes environmental pollution.

An electroplating shop contains probably the most aggressive components of the process industry in as much as they are fluid, corrosive and dynamic. As combatting the effects of these is often quite difficult, great stress is laid on containment through proper selection of materials of construction, sources of energy and process materials. Again the emphasis should be on prevention rather than cure.

The most dangerous combination is corrosion caused by emission of process liquids and gases. These are often variable in composition, concentration, temperature and locality. The materials of containment, capable of withstanding these conditions are often so expensive that they cannot be considered realistically for a low cost environment such as a plating shop. A compromise must often be made relying on various forms of cladding, painting, coating or impregnation. Again a thorough knowledge of characteristics of construction materials greatly assists in selection, economical use and effective maintenance of process equipment. Time and expertise used in selecting proper materials, prevention of ingress of aggressive materials (either liquid or gas) into sensitive working areas of control installations and purging or extracting spent solutions or gases help to maintain equipment efficiently.

PROCESS MATERIALS

Electroplating and other finishing processes involve the required conversion of base surfaces into functional surfaces where the required function is physical or chemical. This conversion occurs mainly by immersion in various alkaline, acid and salt solutions with or without passage of current. Chemical compounds which are used to form the electrolytes are the primary source of danger in the electroplating department and methods must be devised to combat potential hazards from handling these.

Typical dangers emanating from handling chemicals can be chronic or acute, irrespective of the mode of entry and the target organ, and the degree of their insidious effect can usually be described in the form of danger signs, risk phrases and safety phrases. These signs and phrases are now incorporated into the Classification, Packaging and Labelling (CPL) Regulations 1984, which categorise the effects of most of the commonly used chemicals according to their behaviour in animal or material tests. Thus according to the animal trials the effect on skin or internal organs is characterised as very toxic, toxic, corrosive, harmful or irritant, whereas in material trials the effect is determined as explosive, oxidising, extremely flammable, highly flammable or flammable. As a rule one symbol designating the most acute environmental effect and one symbol designating the most acute effect on animal tissue or organ is displayed on the label attached to the receptacle containing the material. In addition, the appropriate risk and safety phrases describing the environmental or toxicological aspects of the material are also displayed. The phrases are standardised in the EEC countries and can be decoded even if only the code *e.g.* R25, S25 is quoted. In addition the substance identification number must be quoted describing the worst ecological or toxicological hazards.

The numbers, symbols and codes do not, however, always help to identify the compound or its concentration. The CPL Regulations make allowances for the nature, concentration or combination of materials so in effect they are designated by the overall hazard represented rather than by the mixture of substances contained.

STORAGE AND PREPARATION OF WORKING SOLUTIONS

Because of the hazardous nature of solutions involved in electrodeposition, consideration must be given to elimination of risk. A few basic rules must be adopted as follows:

1 Do not store acids and alkalis together as excessive quantities of heat or toxic gases might be emitted on mixing.

2 Do not add water to acid as the localised heat evolution might cause 'spitting' or ejection of acid. Always add acid to water.

3 Allow good drainage in the storage area and separate acid drainage from alkaline. The storage area should have mains water and rubber or plastic hose connected to the mains through a tap.

4 Ensure adequate mechanical assistance in materials handling and dispensing facilities in the storage area as containers can be heavy.

5 The first aid box should be available in the storage area or in the immediate vicinity. A First Aider should be on call during mixing or draining and should be trained in procedures required for dealing with possible emergencies, *e.g.* the First Aider should be familiar with the cyanide emergency procedure and with treating hydrofluoric acid burns.

6 Allow adequate ventilation in the storage and dispensing area to prevent concentration of toxic or harmful substances rising above their occupational exposure limit (OEL).

Bibliography

LEGISLATION AND STANDARDS RELEVANT TO SAFE WORKING PRACTICES IN SURFACING
COSHH Regulations 1988
Health & Safety Executive Guidance Note EH 40/88, *Occupational Exposure Limits, 1988,* May 1988
Health & Safety Executive Guidance Note PM 64, *Electrical safety in arc welding,* September 1986
Reporting of Injuries, Diseases and Dangerous Occurrences Regulations 1985 (SI 1985 No 2023)
Health & Safety Executive Guidance Note EH 42, *Monitoring strategies for toxic substances,* 1984
Health & Safety (First Aid) Regulations 1981 (SI 1981 No 917)
Health & Safety Executive Guidance Note MS 15, *Welding,* October 1978
Health & Safety at Work Etc Act 1974
The Factories Act, 1961

British Standards

349 : 1973 Identification of contents of industrial gas containers

349c : 1973 Chart for the identification of the contents of industrial gas containers

638 : Arc welding power sources, equipment and accessories
   Part 6 : 1984 Spec. for safety requirements for construction
   Part 7 : 1984 Spec. for safety requirements for installation and use

679 : 1959 Filters for use during welding and similar industrial operations

697 : 1986 Spec. for rubber gloves for electrical purposes

1542 : 1982 Equipment for eye, face and neck protection against non-ionizing radiation arising during welding and similar operations

1651 : 1986 Spec. for industrial gloves

2092 : 1987 Spec. for eye-protectors for industrial and non-industrial use

2653 : 1955 Protective clothing for welders

2929 : 1957 Superseded by BS 5378

3510 : 1968 A basic symbol to denote the actual or potential presence of ionizing radiation

3664 : 1963 Spec. for film badges for personnel radiation monitoring

4031 : 1966 Spec. for X-ray protective lead glasses

4094 : Recommendation for data on shielding from ionizing radiation
   Part 1 : 1966 Shielding from gamma radiation
   Part 2 : 1971 Shielding from X-radiation

4163 : 1984 Code of practice for health and safety in workshops of schools and similar establishments

4275 : 1974 Recommendations for the selection, use and maintenance of respiratory protective equipment

4513 : 1969 (1981) Spec. for lead bricks for radiation shielding

4803 : 1972 Guide on protection of personnel against hazards from laser radiation

5330 : 1976 Method of test for estimating the risk of hearing handicap due to noise exposure

*British Standards are available from The Sales Department, British Standards Institution, Linford Wood, Milton Keynes MK14 6LE (T:0908 320033) (Tx: 825777)

5345 : 8 parts: 1976-1983
 Code of practice for the selection, installation and maintenance of electrical apparatus for use in potentially explosive atmospheres (other than mining applications or explosive processing and manufacture)

5378 : Safety signs and colours
 Part 1 : 1980 Colour and design
 Part 2 : 1980 Colorimetric and photometric properties of materials

5531 : 1978 Safety in erecting structural frames

5566 : 1978 Recommendations for installed exposure rate meters, warning assemblies and monitors for X- or gamma radiation of energy between 80keV or 3MeV

5741 : 1979 Spec. for pressure regulators used in welding, cutting and related processes

5924 : 1980 Spec. for safety requirements for electrical equipment of machines for resistance welding and allied processes

6158 : 1982 Spec. for safety devices for fuel gases and oxygen or compressed air for welding, cutting and related processes

6691 : Fume from welding and allied processes
 Part 1 : 1986 Guide to methods for the sampling and analysis of particulate matter
 Part 2 : 1986 Guide to methods for the sampling and analysis of gases

6967 : 1988 Glossary of terms for personal eye-protection

Industrial applications of engineering coatings

CHAPTER 10

Use of engineering coatings to improve the wear life of components in industry is worldwide and the circumstances prevailing at each site, which can differ widely, play a significant role in choice of coating process and material. Factors such as working conditions, availability of specialised sub-contract facilities nearby or of on-site skills and equipment and materials are all important, as is a policy of preventative maintenance or reaction to a breakdown.

Previous experience of site conditions is an important factor in deciding the course of action to take and the information in earlier chapters is intended to assist in the decision making process. There is thus no single answer to most problems and the examples given in this chapter must be considered with this in mind. It is fair to say however, that the majority of applications described have developed in circumstances where there is a wide choice of possible solutions. Some readers will have found alternative or better answers to specific problems.

To illustrate the widespread use of engineering coatings, applications have been grouped under a series of industrial headings, with a selection of typical examples under each heading. The lists are not intended to be exhaustive and new applications are being developed continuously. Many industries share similar problems and a study of how other industries have solved their difficulties can often provide the practising engineer with ideas to apply in his/her own sphere of activity.

The coating materials referred to in the lists of applications are those shown in Table 0.1 of the Introduction.

THE AIRCRAFT INDUSTRY

Air travel is a competitive business in which running costs must be tightly controlled while safety and reliability must not be compromised. Military aircraft demand high reliability and exceptional performance and the widespread use of engineering coatings in both types of aircraft including their engines is a good indication of their importance and usefulness, Fig.10.1 and 10.2.

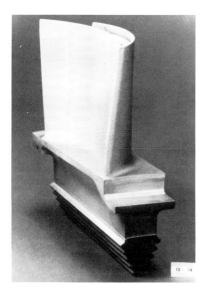

10.1 *High temperature corrosion protection on turbine blade (courtesy Plasma-Technik AG).*

10.2 *Plasma spraying a multilayer thermal barrier coating on an RB 211 nozzle guide vane (courtesy BAJ Coatings Division).*

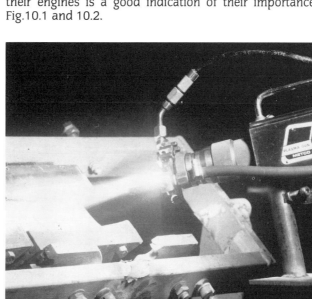

The engine is clearly a major part of the aircraft and application of coatings to engine parts has been practised for over 50 years. In the early days of piston engines, exhaust valves were (and still are) weld surfaced on their seat faces and often all over the crown to maximise life and reliability when working at high temperatures and in the presence of corrosion by the products of burnt fuel.

The special demands of gas turbine engines have provided a great opportunity for surface coatings as the combination of properties required, high temperature strength/corrosion resistance/bearing properties, *etc*, are often impossible to provide from a single material. Many engine parts are fabricated from thin sections of material which would present problems of distortion if welded. This has led to the extensive use of thermal spraying, a process which as noted earlier can also apply non-metallic coatings and has allowed application of thermal barrier materials which effectively reduce the operating temperature of the metallic alloy core of the part.

The fact that thermal spraying can be carried out at low temperatures (compared with welding) means that optimum properties are retained in the substrate material and that reprocessing of the coating can often be carried out after a period in service.

Electroplating is used to protect many parts against atmospheric corrosion and thermally sprayed plastics also find application for the same purpose.

Typical applications

Component	Function	Surfacing material
Engines		
Intake ducts, engine seals	Sliding, fretting wear	Group 6 type 6
Compressor blade contact faces	Impact, fretting wear	6 — 6
Combustion chamber	Thermal barrier	7 — 6
Turbine blades	Hot corrosion	2 — 11
Turbine blade shrouds	Fretting wear	6 — 7
Nozzle guide vanes	Thermal barrier	Multilayer
Knife edge seals	Abrasive wear	6 — 6
Airframes		
De-icing heaters	Electrical resistance	1 — 4
Flap and slat tracks	Abrasive wear	6 — 6
Rivets	Atmospheric corrosion	8 — 2
Ground installations	Atmospheric corrosion	1 — 5

CHEMICAL AND PETROLEUM INDUSTRIES

The chemical and petroleum industries cover a wide range of activities, from extraction of raw materials to production of highly refined end products. Thus all forms of wear are found with special emphasis on corrosion, erosion and also high temperatures in some applications.

To meet these conditions of service special materials are required, which if used for one piece construction of components would involve high costs. Surfacing, using appropriate materials applied to the exposed areas of cheaper substances, enables costs to be minimised, life to be increased and often allows resurfacing to be carried out after a period of use, Fig.10.3-10.5.

Selection of surfacing materials may need careful study; for example small changes in operating temperature or in the composition of the chemical environment can have a significant effect on component life. Certain points need to be kept in mind, for example:

1. Corrosion — assuming the selected coating material is considered to have satisfactory corrosion resistance in the medium and at the temperature involved, the question may arise as to its electrode potential compatibility with the substrate and any other constructional materials. Porosity in thermally sprayed coatings may need to be sealed.

2. Abrasion — materials which depend on formation of surface films to minimise the rate of attack perform badly if this film is continuously removed by abrasion or erosion.

3. Adhesive wear — some materials of high corrosion resistance such as austenitic steels are prone to adhesive wear and 'pick-up' or seizure can easily occur when rubbing against themselves and similar materials.

4. Temperature — the expansion coefficients of the materials used may need to be watched, because of the big differences between some

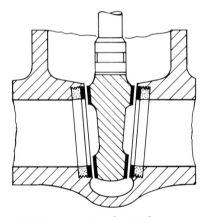

10.3 *Cross section of typical gate valve, showing welded deposits of Group 3 alloys on sealing faces (courtesy Deloro Stellite Ltd).*

10.4 *Surfaces of gate valves used on North Sea oil rigs, coated with tungsten carbide to resist abrasive wear (courtesy Union Carbide Ltd).*

10.5 Centrifugal separator screw surfaced with Group 3 alloys to resist abrasion and corrosion on the diameter and faces of the flights (courtesy Deloro Stellite Ltd).

substrate materials and some coatings. Possible problems to arise include cracking of the surface coating, distortion in use (especially if used at high temperatures) and loss of running clearances.

Typical applications

These examples are grouped according to the principal wear factor involved.

Component	Surfacing material
Abrasion	
Oil well rock drill bits	Group 6 type 4
Oil well rock drill bit bearings	3 — 3
Drill collars	6 — 4
Mud pump sleeves, rods, plungers	2 — 5
Tool joints	6 — 4
Stabilisers	6 — 4 or WC pads in 2 — 5 matrix
Centrifugal separator screw flights	6 — 2
Conveyor screws	6 — 2, 6 — 4
Divertor spools	6 — 2
Adhesive wear	
Mixer seals	3 — 2
Extrusion screw flights	3 — 2, 6 — 6
Corrosion	
Oil storage tanks, structural steelwork	5 — 1, 5 — 2, plastics
Crude oil heat exchangers	8 — 6
Tank heating coils	8 — 2
Thermometer pockets	1 — 5, 2 — 5, 3 — 7
Pipeline and flow control valves and valve trim	2 — 4, 2 — 5, 2 — 12, 3 — 1, 3 — 2, 6 — 6
Erosion	
Valves	As above
Pump impellers	1 — 8, 1 — 9
Pump shafts and sleeves	3 — 7, 6 — 6, 7 — 3
Pipe elbows and bends	1 — 8, 2 — 5

EARTHMOVING, AGRICULTURAL, QUARRYING AND MINING

Earthmoving, agricultural, quarrying and mining industries use plant which suffers primarily from abrasive wear combined with varying degrees of impact, Fig.10.6 and 10.7. The abrasive wear may involve high or low stress contact with the abrasive material.

Alloys used for hardfacing to resist this type of wear rely on the hardness of the metallurgical phases present in their structures and in general, increasing hardness confers increasing resistance to abrasive wear. However, resistance to impact tends to decrease with increasing hardness and in the extreme can lead to cracking or chipping of the surface coating.

Hardness acquired as a result of work hardening through impact — as with austenitic manganese steels — does not lead to the same loss of toughness as the raised hardness may imply, hence its extensive use on parts exposed to abrasion and impact loading.

Deposits of alloys of the harder austenitic matrix type may be subject to contraction cracking (relief cracking) if applied in thick layers or on large areas. Where extensive building up is required to restore worn areas, it is normal to restore the profile with a deposit of a softer and tougher low alloy steel, which limits the amount of hardfacing to be subsequently deposited but provides adequate support.

Table 10.1 shows the general relationship between abrasion resistance and toughness/resistance to impact of the most important groups of materials used in these industries. There is of course some overlapping between the various alloys in each group and between groups.

10.6 *Excavator bucket for limestone hardsurfaced by MMA using alloy steel deposit (courtesy ESAB Group (UK) Ltd).*

10.7 *Extensive use of engineering coatings extends the life of the toothed buckets on this excavation unit (courtesy ESAB Group (UK) Ltd).*

Table 10.1 Relationship between abrasion resistance and toughness/resistance to impact

Group	Abrasion resistance increasing	Impact resistance increasing	Typical uses
6 Tungsten carbide materials		↑	Sinter plant, rock drills, brick and clay tools
2/5 Hi-Cr complex irons			Hot wear — sinter plant scrapers and screens
2/4 Hi-Cr martensitic irons			Ball mill liners
2/3 Hi-Cr austenitic irons			Shovel teeth, screen plates, bucket lips
2/2 Martensitic irons			Abrasion and adhesive wear
2/1 Austenitic irons			Crushers, pump casings and impellers
1/7 Austenitic Cr-Mn steel			Heavy impact e.g. excavator buckets
1/6 Austenitic Mn steels			Heavy impact e.g. crushers, hammers
1/2 Low alloy steels			Build-up materials, tractor parts
1/1 Carbon steels	↓		Build-up materials

The presence of impact calls for a coating which has the strongest possible bond with its substrate and which can provide protection in depth. In consequence, such applications usually call for welded deposits. Thermally sprayed deposits find application on some bearing surfaces not normally exposed directly to impact by the minerals being handled.

The alloy selected and also the pattern of weld beads applied to tools depend on working conditions. Cost effective deposits do not necessarily need to provide all-over coverage; for example, a 'waffle' or criss-cross pattern is used when handling wet sand or clay, which packs in between the beads and protects the substrate. A 'dot' pattern is used on difficult to weld substrates when overheating must be avoided. When handling rock a series of beads running parallel to the flow of material causes the spoil to rub on them and not on the substrate.

Typical applications

Component	Surfacing material
Tractor parts, *e.g.* rollers, idlers, rails, sprockets, *etc*	Group 1 type 2
Tractor mounted tools *e.g.* ripper teeth, bulldozer blades, shovel buckets, teeth, *etc*	2 − 4, 2 − 6, 1 − 6, 1 − 7, 2 − 1, 2 − 2, 2 − 3
Crushers, crusher rolls	1 − 6, 1 − 7, 2 − 1
Screens	2 − 3, 2 − 4
Hammers	2 − 4
Cones and bowls	2 − 4
Jaws	1 − 6, 2 − 4
Hydraulic rams	Hard Cr plate

INTERNAL COMBUSTION (PISTON) ENGINES

The ubiquitous piston engine fitted with poppet valves depends on an effective gas tight seal when the valves are shut to develop its rated power output.

The exhaust valve, surrounded by the hot products of combustion during the exhaust stroke, rises to high temperatures until it shuts and the heat is dissipated through the seat in the engine and the stem. Precise working conditions depend on engine design, fuel used and operating conditions but the valve seat surface is exposed to high temperatures, corrosion, thermal and mechanical shock. Temperatures are highest in highly rated petrol engines and corrosion is greatest in diesel engines running on residual fuels.

The diameter of the valve head may be 25mm or less on small petrol engines up to 500mm or more on large diesel engines. A variety of steels and some nickel base alloys are used to provide adequate mechanical properties in the valve at operating temperatures, but it is frequently necessary to provide special coatings on the valve seat areas to minimise valve and seat wear, indentation of seat surfaces by hard particles produced in the combustion space and to give the best possible protection against erosive wear from the high velocity exhaust gases.

The end of the valve stem is in contact with the valve actuation device, *e.g.* the tappet or rocker and valve materials chosen for good high temperature strength and corrosion resistance may need local hardening on the stem tip to resist deformation and wear from the forces necessary to open the valve.

For valve seating surfaces in diesel engines it is normal to use welded deposits of Group 2 type 5 or Group 3 alloys, whereas for petrol engines Group 3 type 4 alloy is widely used. Valve stem tips are often weld deposited with Group 3 type 2 or 3 alloys.

The valve stem slides in a lubricated guide bush and electrodeposited hard chromium is normally used on this.

Manufacture of new valves usually incorporates a mechanised welding operation to deposit the surfacing alloy using purpose built machines based on the PTA process — as illustrated in Chapter 2 — or TIG/MIG or less frequently by oxyacetylene torch.

As valve size increases, so does the cost of substrate materials, even when a low alloy stem is friction welded to the head. Thus valves with a crown diameter in excess of about 75mm are regularly restored by resurfacing, an operation often carried out manually in areas remote from more highly developed servicing facilities.

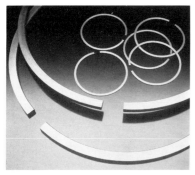

10.8 *Molybdenum coated piston rings (courtesy BAJ Coatings Division).*

Other important applications of surfacing in the internal combustion engine field include coated piston rings (Fig.10.8) and spraying of piston crowns (Fig.10.9).

10.9 *Plasma spraying a piston crown with a thermal barrier coating (courtesy Metco Ltd).*

Typical applications

Component	Surfacing material
Exhaust valve seat area, petrol	Group 3 type 1, 3 — 4
Exhaust valve seat area, diesel	3 — 3, 2 — 5
Valve stem tips	3 — 2
Rocker pads, cross heads	1 — 9, 3 — 2
Valve stems	Hard Cr plate
Piston rings	5 — 5, hard Cr plate
Piston crowns	7 — 1
Bearing shells	Pb-Sn, Pb-Sn-Cu; Pb-In
Diesel cylinder liners	Electrodeposited porous Cr plate

FOOD

Food is essentially from two sources — animal and vegetable farming. However, much of the food consumed by humans and farmed animals is processed and the industry therefore faces problems similar to those of the earthmoving and chemical industries, coupled with the overriding need for hygiene, which can, for example, involve use of corrosive cleaning media.

Farm cultivation of vegetable materials including grain, root and stem crops faces problems of abrasive wear with some impact if soils are stony and the surfacing materials appropriate to protect cultivation machinery are those discussed in the section of this chapter which deals with earthmoving and associated activities.

Many vegetable and dairy products such as beet and cane sugar, milk and cream are processed and the equipment is usually made of stainless steel for reasons of hygiene. The low hardness and low resistance to adhesive wear of this type of steel indicates use of more wear resistant engineering coatings in parts such as pumps and valves, using materials of good wear resistance such as those from Groups 2 and 3, Fig.10.10.

10.10 *Can seaming roll of Group 3 type 1 alloy before coating with TiN by PVD (courtesy Deloro Stellite Ltd).*

Processing of meat, involving cutting or mincing operations, calls for durable, sharp cutting edges of corrosion resistant materials and again, weld surfacing of corrosion resistant substrates with surfacing alloys of Group 1, 2 and 4 provides an effective answer.

Margarine substitutes for butter are produced using vegetable oils. These are extracted from a variety of plants such as seeds and nuts in expeller machines, which comprise a long steel screw running inside a close fitting barrel, rather like a mincing machine and developing pressure which squeezes the oil from the seed. The residue of vegetable matter is used as a basis for cattle cake. To reduce wear on the screw and barrel, the screw flights are coated with wear and corrosion resistant alloys typically of Group 1 or 3 and the barrels are lined with similar alloys by a specialised coating process designed for the purpose.

At the end of the food chain, unused food is frequently comminuted by a waste disposer and flushed down the sewage disposal system. The water borne effluent also carries a proportion of solids including woven materials which must all be comminuted before treatment. To maintain the sharp cutting edges necessary for this type of duty, welded deposits of hard alloys resistant to corrosion are used, typically of Group 2 or 3, depending on the machine part and its working conditions.

Typical applications

Component	Surfacing material
Sugar cane cutter knives	Group 3 type 1, 6 — 4
Shredder hammers	1 — 10, 6 — 4
Grain crusher rolls	Cr steel
Pipeline valves and seats	3 — 1, 3 — 2
Homogeniser valves and seats	3 — 3
Pump shafts, sleeves, impellers	Various
Expeller screws	3 — 2, 1 — 10
Expeller cage bars	3 — 1, 3 — 7
Doctor knives	3 — 2
Can seaming rolls	3 — 3 plus TiN by PVD
Packaging knives	3 — 2
Sewage shredder knives and pumps	3 — 2, 3 — 3
Brewing containers	8 — 6
Food mixer bowls	8 — 2

FORGING

Shaping of components in forgeable alloys is carried out at elevated temperatures by press, hammer or drop forging. This work calls for tools which resist deformation, abrasion by mill scale and cracking caused by thermal and mechanical stress.

As in other examples already referred to, this requires properties on the surface of the tools which are not easy to provide from the materials used for the bulk of the tool. The tools are often quite massive and may be rendered unusable by a small amount of wear on a working face. Weld surfacing with appropriate alloys can prolong the initial life and provide the opportunity to carry out rebuilding of the surfaces when worn.

Selection of coating alloy depends on the working conditions of the tool. Alloys of Groups 2 and 3 are used to provide good high temperature properties, Fig.10.11 and 10.12; for example, a Group 2 type 4 alloy is resistant to thermal and mechanical shock and is used on drop forging dies. It work hardens in use which helps it to retain the tool profile without cracking. Tools used for press forging lack the impact required to develop

10.11 *Milling die cavity deposited with Group 2 type 4 alloy to finished size (courtesy Deloro Stellite Ltd).*

10.12 *Flash trimming die deposited with Group 3 type 1 alloy (courtesy Deloro Stellite Ltd).*

work hardening so a harder alloy which is a modification of the Group 3 type 1 alloy is used.

Cutting devices such as shears and clipping beds require a still harder alloy and a Group 3 type 1 alloy is used.

Typical applications

Component	Surfacing material
Drop forging die impressions	Group 2 type 4
Drop forging die flash lands	2 — 4
Press forging die impressions	3 special
Clipping beds	3 — 1
Hot shears	2 — 12, 3 — 1
Rotary forging hammers	3 — 2
Furnace skid rails	3 special

GLASS

Production of hollow glasswear such as drinks bottles, jars and other containers uses a series of operations to mould or shape a 'gob' of glass which is sufficiently hot to be plastic. At the final shaping stage, the tool in contact must extract sufficient heat from the glass to ensure that it retains its shape when ejected.

Tools used include split moulds to form the outside of the container and plungers to develop the internal form. Other tools simultaneously form the shape of the neck and profiles to engage with the lid or other closure.

Tools are typically made of cast iron or bronze which gives higher thermal conductivity. When assembled to form a complete mould there are a number of joints between the individual parts; when new, there is no visual evidence of these on the formed glass, but after a period of use, corners become worn or even chipped and this shows up on the finished glass container.

To obtain the longest life of the tools and to enable parts which ultimately do wear to be refurbished, a technique was developing using powder welding as described in Chapter 2. Worn corners are first chamfered to provide a small flat platform, perhaps 3mm wide, which is then built up with a self-fluxing alloy of Group 2 type 5. One of the softer alloys of this family is used, enabling the original sharp corner to be restored by a hand operation such as filing.

Glass plungers used for piercing the glass gob are surfaced with a similar alloy, but the spray fused process is more suited to the coverage of larger areas than powder welding and has the advantage of giving smoother deposits on such parts which require less work to finish to size, Fig.10.13.

10.13 *Glass plungers coated with Group 2 type 5 alloy 40RC by spray fusing (courtesy Deloro Stellite Ltd).*

Typical applications

Component	Surfacing material
Press/blow plungers	Group 2 type 5
Blow-blow plungers	2 — 5
Blank and blow moulds	2 — 5
Guide plates and rings	2 — 5
Neck rings	2 — 5
Take-out jaws	2 — 5, 6 — 2

PULP AND PAPER

Paper is manufactured in a large number of qualities using variations of a basic production process. It may be made from raw materials such as wood fibres or contain or consist of recycled waste material. The process involves removal of bark from tree trunks, breaking down the wood into chips, forming pulp mechanically/chemically and cleaning/modifying the fibrous structure of the pulp. It is then converted into paper or board and cut to size.

Viewed as a whole the processes involve wear of all the types described in Chapter 1 and these occur in various combinations. Corrosion of plant occurs at several stages of manufacture especially in chemical pulping and bleaching. It would not be unusual for normal materials of plant construction to have a short life in these circumstances and long term protection is achieved by use of appropriate engineering coatings, Fig.10.14-10.16.

10.14 *Rotary feed valve for digester manually deposited with Group 3 type 2 alloy before machining to size (courtesy Deloro Stellite Ltd).*

10.15 *Paper mill cylinder, 11m long and 1.5m diameter metal sprayed with a mixed bond coat of 1-1 and 4-3 materials and a final layer of mixed 1-3 and 1-5 materials (courtesy Metallisation Ltd).*

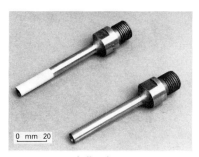

10.16 *Paper drills plasma arc sprayed with Group 6 type 6 material to a thickness of around 0.05mm (courtesy CSIRO, Australia).*

Use of such protection is not limited to paper manufacture, there are applications in cutting, folding, gluing and printing of paper and board which increase life and reduce downtime.

Typical applications

Component	Surfacing material
Debarking flails	Group 1 type 10, 6 — 4
Chipper knives	3 — 2
Chipper wear plates	3 — 1, 3 — 2
Conveyor and feed screws	1 — 10, 3 — 2
Pulper segments	1 — 10, 3 — 3
Pulper drive shafts	3 — 7
Dirt trap cones	2 — 5
Rotary feed valves	3 — 2
Shaft sleeves for finers, defibrators	3 — 7
Disintegrator hammers	3 — 3, 6 — 4
Packaging knives	3 — 2

PLASTICS AND RUBBER

In the plastics and rubber industries a number of different operations are carried out, each of which has its own group of wear problems. They have been grouped together as there are similarities in certain operations which are employed in the two industries.

Both use extrusion machines which are similar to the extruders mentioned in the section on the food industry, although there are significant differences in the design of the screw flights. Protection of the crest of the flights is carried out by weld surfacing using alloys from Group 2 and 3 applied by manual and mechanised welding processes, Fig.10.17. Thermally sprayed deposits of Group 6 type 6 are sometimes used. Barrel liners are coated by a spin casting process.

Plastic materials are produced in a semi-finished form as pellets from pelletiser plates and cutters whose cutting edges are coated with a welded or thermally sprayed deposit of a Group 6 type 6 material, using D gun or Jet Kote processes, Fig.10.18.

Mixers of the Banbury type suffer wear of the internal surfaces of the body and on the rotors; these and the rotor seals are typically protected with

10.17 *Extrusion screw for plastics weld deposited on the flight diameters before regrinding to size (courtesy Deloro Stellite Ltd).*

welded deposits of Group 3 alloys applied by MMA, MIG/MAG or SAW processes, Fig.10.19.

A difficult application is protection of biffer screens used in polymer production. The screens are perforated metal rings ~760mm diameter inside which rotates a beater cross. Glass filled polymer granules pass through the screen causing wear. A solution to the wear shown in Fig.10.20 on an unprotected type 321 stainless steel screen has been found by applying a coating of 55% tungsten carbide in a Group 2 type 5 matrix by the spray fuse process. Figure 10.20b shows the wear after 21 days use following application of the coating. The main problem in surfacing is control of distortion.

10.18 *Pelletiser plate deposited with tungsten carbide (courtesy Union Carbide).*

10.19 *Banbury mixer components with welded deposit (courtesy Deloro Stellite Ltd).*

10.20 *Biffer screen handling glass filled polymer: a) After two days' service — unprotected stainless steel; b) After 21 days' service following coating with Group 2 type 2 material (courtesy ICI).*

Typical applications

Component	Surfacing material
Extrusion and injection moulding screw flights	Group 1 — 8, 2 — 5, 3 — 2, 6 — 6
Rotary chopper blades	3 — 2
Pelletiser plates and cutters	3 — 2, 6 — 6
Banbury mixer rotors, bodies, doors and end plates	3 — 2, 3 — 3

POWER GENERATION

Generation of electricity by thermal means accounts for a number of interesting uses for engineering coatings. Mention has already been made of such uses in gas turbine and diesel engines and this section is concerned with steam turbines coupled to fossil or nuclear fuel heat sources.

Coal fuel is widely used and for large boilers it needs to be pulverised, creating abrasive wear of the pulverisers. When burned, products of combustion cause hot corrosion and the ash content causes abrasion of boiler tubes and induced draught fans.

Nuclear fuel creates its own problems for mechanisms which control the movement of fuel rods, *etc*, in the reactor coolant, which are subject to adhesive wear.

Steam generated in the boiler/heat exchanger is admitted to the turbines through various control valves which operate at high temperature and can suffer erosion of seating surfaces.

Typical applications

Component	Surfacing material
Coal treatment plant	Group 1 type 10, 6 — 4
Boiler tubes	50:50 Ni-Cr
Induced draught fan blades	6 — 4
Hydroelectric turbine blades	1 — 3
Turbine stop, throttle, intercept valve seats	3 — 1 or Co-Cr-Mo type
Valve grinding surfaces	3 — 1, 3 — 7
Steam pipe expansion joints	3 — 1

STEELMAKING

The operations required to convert iron ore into rolled steel products cause rapid wear of production plant, much of it due to the high processing temperatures used and the abrasive effects of mill scale.

Use of engineering coatings has been practised for many years as a means of prolonging life between breakdowns and allowing repeated repairs to large and expensive items of plant, Fig.10.21 and 10.22. Many of the applications

10.21 *Hot mill roll neck reclaimed with a thermally sprayed deposit of Group 1 type 2 steel up to 25mm thickness (courtesy Metallisation Ltd).*

10.22 *Twin high velocity thermal spray Jet Kote guns applying a deposit of Group 6 type 6 material to the surface of a steelworks pickling line roll to a thickness of 0.08-1.5mm and to a desired roughness between Ra3 and Ra12. Life improvements over an uncoated roll is about 4:1.*

call for thick coatings resistant to thermal and/or mechanical stress and weld deposited coatings are usually best suited to this type of work. Arc welding is widely used because of the heavy sections of components to be built up.

Typical applications

Component	Surfacing material
Coal crushers	Group 1 type 10, 6 — 4
Ore crushers	1 — 6, 1 — 8
Shovel teeth, bucket lips	1 — 10, 1 — 8
Sinter plant high wear areas	1 — 12, 6 — 4
Furnace conveyor skidder bars	3 — 1
Bell and hopper seals	2 — 4
Explosion valves	3 — 1
Ingot lifting dogs	1 — 4, 3 — 1
Hot slab shears	2 — 4, 3 — 1
Roll neck abutment rings	3 — 2
Table rolls	1 — 9
Roll journals	1 — 3
Down coiler rolls	1 — 3, 2 — 4
Bar mill entry guide rolls	3 — 2
Acid pump parts	3 — 2
Galvanizing bath rolls, bushes	3 — 1
Pickle line rolls	6 — 6
Tube piercer points	2 — 4, 3 — 1
Rail points and crossings	1 — 2, 1 — 6, 1 — 7
Crane brake drums	6 — 6
Roll bearing seals	7 — 3
Plating line seals	5 — 3

TEXTILES

Both natural and synthetic fibres are used in textile production, which is carried out on high speed machines and these speeds lead to rapid wear caused by abrasion from the thread or adhesive wear between machine parts. Use of conventional lubricants is often not possible, because of the risk of staining the product, so surface coatings having good bearing properties can provide an ideal solution, Fig.10.23.

Corrosion is not normally a problem except in manufacture of artificial fibres which require coagulation in a corrosive solution. The continuous fibre is cut to form tow of a staple length in the coagulating solution and, for this purpose, knives must retain a sharp cutting edge by use of a corrosion resistant and hard alloy. For this, alloys of Group 3 type 2 are used.

10.23 *Thread overrun roller and thread guides coated with ceramic by plasma spraying to protect against frictional wear (courtesy Plasma-Technik AG).*

Typical applications

Component	Surfacing material
Thread guides, overrun rollers	Group 7 type 1, 7 — 3
Machine cams	3 — 3
Looper and latch needles	3 — 3
Winding spindles	6 — 6
Flattening rolls	6 — 6
Crimper rolls	3 — 2
Carpet trimming knives	3 — 2

Some applications lend themselves to use of alloys of the 3 — 3 type in the form of precast inserts, brazed in position, rather than surfacing.

TIMBER

Growing timber, particularly fast growing tropical hardwoods, are often contaminated with abrasive mineral matter such as sand or tramp metal such as wire, nails or bullets. Tree trunks are usually cut with band saws running at high speed, the teeth of which need to have high resistance to abrasive wear and frictional heat.

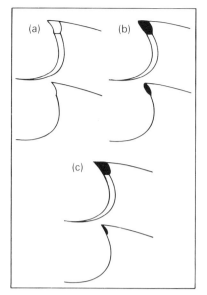

10.24 Hardfacing of saw teeth:
a) Each tooth is swaged; b) The dimple is filled by hardfacing and the tooth tempered; c) Tooth is ground until alloy front is correctly faced and sharp.

Until the technique of tipping band saw teeth by weld surfacing was developed some years ago, saw mills tended to accumulate tree trunks which could not be cut economically with normal steel saws (high carbon steel). Hardfacing was originally carried out using an oxyacetylene torch and this is still practised. Special machines have also been developed to do the job automatically using arc welding methods.

The sequence of operations is shown diagrammatically in Fig.10.24, the swaging operation being the same as that used in normal resharpening to give side clearance to the tooth. As the saw band is a high carbon steel and therefore tends to harden under the weld, a light tempering operation is carried out before each tooth is ground to give the usual cutting edge and angles. Other applications of engineering coatings to equipment handling wood are covered in the section of this chapter dealing with pulp and paper.

Typical applications

Component	Surfacing material
Veneer pressure bars	Group 3 type 2
Band saw teeth	3 — 2

TRANSPORT
Use of engineering coatings in the transport industry falls into three categories — power units, other vehicle components such as suspensions and brakes and fixed permanent structures such as bridges.

The working environment may be sea, land or air and the latter is dealt with in the section on aircraft. Piston engines are covered in their own section and here we are concerned with road, rail and marine transport.

Road transport — vehicle components
Those areas of road vehicles exposed to the abrasive effects of grit and corrosion caused by salt can show severe deterioration in a short period of use. Manufacturers are increasingly improving the life of these areas by use of zinc coatings and wax injection, but these have limited resistance to abrasion. Thermally sprayed coatings of epoxy based plastics are now applied to exposed areas such as wheel arches, bumpers and spoilers and similar coatings are used as body coats on some vehicles. Chassis of heavy transport vehicles are typically sprayed with zinc while worn brake drums are reclaimed with arc sprayed deposits of Cr-Mo steel, Fig.10.25 and 10.26.

10.25 Semi-trailer flame sprayed with zinc to a depth of about 0.1mm, to be followed by the normal decorative paint finish, giving a life of about ten years without maintenance (courtesy Metallisation Ltd).

10.26 Reclamation of a passenger bus brake drum by arc spraying an initial coating of molybdenum, followed by a working layer of low alloy steel (courtesy Metallisation Ltd).

Railways — vehicle components
Coach coupler knuckles are protected from new with thermally sprayed and fused deposits of tungsten carbide in a matrix of a self-fluxing alloy of Group 2 type 5, to resist dry rubbing and some impact, Fig.10.27. Other parts successfully reclaimed include brake torque reaction tubes plasma sprayed

with tungsten carbide/cobalt of Group 6 type 6, and brake pad actuator pivots sprayed and fused with a Group 2 type 5 alloy.

Location surfaces for ball and roller journal bearings which suffer from fretting wear are reclaimed by thermal spraying; Fig.10.28 shows a traction motor armature spindle set up for arc spraying with a 13%Cr steel of Group 1 type 3.

Arc spraying of copper on the surface of coach axles is used to create a brush path. The deposit is subsequently rolled to compact the deposit.

Railways — track
Points and crossings are subject to greater wear than track and are repaired with arc welded deposits of Group 1 type 2, 6 or 7 materials.

Tools used to remove the hot flash from butt welded rails are surfaced with a Group 3 type 2 alloy.

Marine transport
The environment of salt water, often with abrasive sand, shells, *etc*, in suspension poses combined wear and corrosion problems for exposed equipment. Typical examples of the use of engineering coatings on such parts are the bearing surfaces of rudders, stabilisers and submarine hydroplanes. These bearing surfaces, which can be up to about 600mm diameter are weld surfaced with an alloy similar to Group 3 type 1 alloy, giving a substantial improvement in life and avoiding regular and costly dry docking for inspection and possible replacement, Fig.10.29.

The operating gear of fishing vessels suffers rapid corrosion and wear; sprayed and fused coatings of Group 2 type 5 and Group 3 type 7 alloys find application on bearing surfaces while protection against seawater corrosion of other parts can be achieved with sprayed deposits of plastics.

Two interesting onboard ship reclamation jobs are the coating of worn diesel engine piston skirts (~ 300mm diameter) with aluminium bronze of Group 4 type 3, and of the worn skirt of a propellor shaft bearing, arc sprayed with Monel metal.

Typical applications

Component	Surfacing material
Marine main drive shafts	Group 2 type 2
Submarine hatch covers	2 — 12

10.27 *Manufacture of high speed train couplings incorporates coatings on the rubbing surfaces exposed to unlubricated wear and impact. The coating is a thermally sprayed layer of Ni-Cr-Si-B alloy containing tungsten carbide which is then furnace fused. Lifetime is raised typically from three months to six years or more (courtesy British Rail).*

10.28 *Rail traction motor shaft bearing seating reclaimed with a gas or arc sprayed deposit of Group 1 type 3 steel (13%Cr), before machining to size (courtesy British Rail).*

10.29 *Marine shaft (courtesy MoD and Deloro Stellite Ltd).*

Glossary of terms used in surfacing

CHAPTER 11

The following is a list of the more specific terms used in this handbook. Reference should be made to appropriate standards for the definition of other engineering terms.

Abrasive wear — loss of metal from a surface by the mechanical cutting action (rubbing or friction) of a secondary material which is in relative motion to that surface.

Adhesive wear — wear brought about by welding under pressure of surface asperities and subsequent shearing of such welds to give surface damage.

Anodising — production of an oxide film on a metal surface, chiefly aluminium, by making it the anode in an electrically conducting solution.

Arc spraying — a thermal wire spraying process in which the heat source is an electric arc; the molten metal is projected from the arc by means of a gas jet.

Bond — adhesion between coating and substrate; the type of bonding mechanism (e.g. mechanical or metallurgical) and the bond strength depend on the process and materials used.

Bond coat — a thermally sprayed coating applied to provide a 'key' for subsequent build-up coats of a different material (see also exothermic coating materials).

Bond strength — stress required to detach the coating from the substrate.

Building up — the process of restoring a worn or damaged region of a component by deposition of material until the required dimensions are obtained.

Buttering — the technique of depositing an intermediate layer (or layers) of weld metal between the coating material and the substrate.

Cavitation erosion — a form of erosion causing material removal because of the collapse of vapour bubbles at the liquid metal interface in a turbulent flowing liquid.

Cermet — a composite material comprising ceramic and metal constituents.

Chemical vapour deposition (CVD) — the production of a metallic or metal compound coating by chemical reaction or thermal decomposition of a gas; reaction usually takes place at temperatures exceeding 700°C on the substrate surface.

Cladding — a term used loosely to describe the deposition of a thick coating (~1mm) on to a relatively large area.

Coating — a deposited layer of material which adheres to the substrate. The material in thermal spraying or welding from which the coating is derived.

Composite plating — electrodeposition of a metal plus a non-metal, typically a carbide or oxide, to form a wear resistant coating. The metal is in solution as a salt but the non-metallic component is present as a suspension of particles that become entrapped in the deposit.

Consumables — all materials used in production of a coating including wire, powder, flux, rods, electrodes, gases and materials for surface preparation.

Contact fatigue (surface fatigue) — failure of surfaces in rolling contact subject to fatigue loading.

Cord — a continuous consumable for thermal spraying comprising the surfacing material contained within a plastic sheath.

Delamination wear — wear in rolling sliding surfaces, believed to involve a fatigue mechanism, in which voids or cracks below the surface coalesce to allow sheets of the coating to detach.

Deposit — see coating.

Detonation spraying — a thermal spraying process in which the controlled explosion of a mixture of fuel gas and oxygen is used to melt and propel a coating powder to the substrate.

Dilution — a mixing of the deposit with the surface zone of the substrate, build-up or buttering layers during the deposition process.

Electroless plating — an aqueous, usually catalytic, reduction process in which metal is deposited, without the passage of a current, from a solution of its salts.

Electrophoretic deposition — a technique in which the coating material is in the form of positively or negatively charged colloidal particles often suspended in a non-ionising liquid. Particles drift under an applied voltage to form coatings at the appropriate electrode.

Electroplating — the deposition of a layer of metal on a cathode immersed in a conducting solution. The solution contains a salt of the metal to be deposited.

Erosion — mechanical removal of material from a surface caused by the flow of a liquid or gas containing particulate matter or liquid droplets.

Evaporation (vacuum) coating — a method of coating an object, held in a vacuum, by condensing the coating-material vapour on to it. The coating vapour comes from a heated source which is usually of metal.

Exothermic coating material — a material which undergoes an exothermic chemical reaction when heated in a thermal spray gun; the particles reach a higher temperature than that achieved with conventional coating materials which improves their adhesion to the substrate (most often used as bond coats).

Flame spraying — a thermal spraying process in which the heat source is an oxyfuel gas flame. The molten metal is projected by a jet of compressed air or gas.

Flash evaporation — a vacuum evaporation method of applying an alloy coating by sprinkling the powdered alloy on to a flat crucible held at a very high temperature. The temperature is sufficiently high to cause all constituents of the alloy to evaporate effectively at the same rate — this does not occur during conventional vacuum evaporation.

Flux — material used in welding to remove surface oxides, protect the substrate and weld metal from reactions with atmospheric gases, and assist in shaping the weld deposit. In arc welding processes the flux also protects the arc zone from atmospheric reactions and may be used to transfer alloying elements to the molten pool.

Finishing — any process applied to an as-deposited coating to produce the desired quality of surface finish.

Fretting — surface damage occurring between two surfaces in close contact (usually when under heavy load) and when subject to small relative movement.

Fused coating — a thermally sprayed coating fused by the application of heat, *e.g.* by gas torch, induction, furnace, laser, *etc.*

Galling — damage to one or both surfaces of materials sliding in contact with each other caused by local welding of high spots.

Graded coating — a system in which a succession of layers is applied, each successive layer being of a higher concentration of the material of the final coat. Used to minimise dilution effects when the composition of the substrate and coating are widely different.

Grit blasting — a surface preparation process which uses hard angular particles projected at the substrate to provide a rough surface for the subsequent reception of a sprayed coating.

Hardfacing — the application of a hard, wear resistant coating by welding or spraying.

Heat affected zone — the zone underlying a weld deposit in which the metallurgical state of the substrate material has been changed by heat input from the welding operation.

Impact damage — mechanical damage caused by shock loading.

Interpass temperature — the temperature attained by the substrate or previously deposited weld metal immediately before deposition of a subsequent weld run.

Ion implantation — this is a surface modification, rather than a coating, technique. A very high energy beam of, e.g. ionised nitrogen, is fired at a metal object in vacuum and penetrates a short distance (typically $0.1\mu m$) below the surface. The wear resistance of such surfaces can be superior to those conventionally nitrided.

Ion plating — an evaporation coating technique that takes place in an ionised inert gas at a pressure of about 10^{-2} torr. The substrate is made the cathode which is biased at a few kV negative and is ion bombarded in the resulting glow discharge.

Line of sight coating — coating processes in which the coating material travels in a straight line from the source to the substrate; they cannot coat around the back of an object, nor into re-entrants or on to any area shadowed from the source.

MAG welding (metal active gas) — gas shielded metal arc welding using a consumable wire electrode where the shielding is provided by a shroud of active or non-inert gas or mixture of gases.

Manual metal arc welding — metal arc welding with straight covered electrodes of a suitable length and applied by the operator without automatic or semi-automatic means of replacement. No protection in the form of a gas or mixture of gases from a separate source is applied to the arc or molten pool during welding.

Mechanical bond — adhesion between the coating and substrate achieved principally by interlocking of the particles of the coating with specially prepared asperities on the surface of the substrate.

Mechanical plating — a non-electrolytic method which consists of barrel coating components in a colloidal suspension of the depositing metal. Reaction between the objects and the particles causes them to coalesce and to bond to the surfaces of the components.

Metallurgical bond — adhesion between the coating and substrate achieved by diffusion across the interface between coating and substrate.

MIG welding (metal inert gas) — gas shielded metal arc welding using a consumable wire electrode where the shielding is provided by a shroud of inert gas.

Overlay coating — surface coating that lies substantially on top of the substrate.

Physical vapour deposition (PVD) — a generic term that includes evaporation coating, ion plating, sputtering, and — somewhat peripherally — ion implantation. PVD processes are those in which the coating material is transported from source to substrate as an elemental vapour, unlike CVD in which transport is in the form of a chemical compound.

Penetration — the depth by which the fused zone created by a weld deposit extends into the substrate or previous weld deposits, see also dilution.

Plasma arc welding — an arc welding process in which the heat for welding is produced with a constricted arc between: (a) an electrode and the substrate (transferred arc) or (b) between an electrode and the constricting nozzle (a non-transferred arc). Shielding is provided by the hot ionised gas issuing from the orifice and an auxiliary shroud of inert gas.

Plasma spraying — a thermal spraying process in which the heat source is a high temperature stream of partially ionised gas produced in an electric arc.

Post-heat treatment — the application of thermal treatment following completion of the surfacing process.

Powder spraying — thermal spraying using a coating material in powder form.

Powder welding — a surfacing process in which the coating alloy in the form of a powder is fed into an oxyfuel gas welding torch from which it emerges through the flame to give a fused deposit.

Preheat treatment — the application of heat before and during surfacing.

Pyrolytic deposition — this term is generally confined to CVD reactions which take place by thermal decomposition on very hot substrates and which produce very dense coatings.

Rod spraying — thermal spraying using a coating material in rod form.

Running-in — an initial period during the operation of new mechanical equipment in which adhesive wear damage may occur to metallic components.

Scuffing — form of severe adhesive wear, most often observed during running-in, which involves significant transfer of material between the surfaces.

Sealant — material used for sealing a porous coating.

Sealing — the impregnation of a porous coating with material to seal the pores.

Self-fluxing alloy — a surfacing alloy in which elements such as boron and silicon are present to react with oxides during a fusing operation to give a low melting point slag.

Slag — a fused mixture of flux and impurities which floats on the surface of the molten pool.

Spalling — detachment of pieces of coating resulting from the continued application of compressive loads or thermal stress.

Spray fuse — a process in which material is first applied to a prepared surface by thermal spraying and then fused by application of additional heat.

Sputter coating — the coating material is in the form of the cathode in a low pressure glow discharge system. Bombardment of the cathode by ionised low pressure gas causes atoms to be ejected from the cathode surface, *i.e.* 'sputtered' and deposits target atoms on nearby surfaces which are 'sputter coated'.

Stress relief — a thermal treatment applied to a structure or component to reduce the magnitude of residual stress.

Submerged-arc welding — metal arc welding in which a bare wire electrode or electrodes are used; the arc or arcs are enveloped in a granular flux, some of which fuses to form a removable covering of slag on the weld.

Substrate — the parent material of the component to which a coating is applied.

Surfacing (surface coating) — the application of a coating to a component to change or restore its surface behaviour.

Surface engineering — the technology involved in understanding and controlling the properties and economies of engineering surfaces.

Surface preparation — cleaning and/or roughening the substrate surface ready for application of a coating.

Thermal spraying — a process in which a heat source raises the coating material to plastic or molten state and projects it in finely divided form on to the substrate.

Throwing power — a measure of the ability of a coating process to deposit material uniformly on to complex surfaces or on to objects that are not conformable to the source.

TIG welding (tungsten inert gas) — gas shielded arc welding using a non-consumable pure or activated tungsten electrode where the shielding is provided by an inert gas shroud.

Tubular wire — a surfacing wire made by the inclusion of alloying elements, fluxes, *etc*, in a sheath made from a ductile metal or alloy such that the total composition of the wire is that desired.

Ultimate resilience — defined as $\frac{(\text{ultimate tensile strength})^2}{2 \times \text{modulus of elasticity}}$ and claimed to indicate resistance to erosive wear conditions.

Wear — progressive loss of material from the operating surface of a body because of relative motion between the body and another with which it is in contact.

Wear coefficient — term in Archard wear equation, nominally a constant under usual (industrial) operating conditions.

Wire spraying — thermal spraying using coating material in wire form.

Sponsoring companies

Coated Electrodes UK Ltd, Celcoat Division — a range of sub-contract thermal spraying services including plasma, oxy/gas powder, rod and wire as well as fused nickel or cobalt alloy coatings. Technical facilities are available for coating development and analysis as well as machining and finishing services. Coated Electrodes offers particular expertise in machining and coating graphite products.
Coated Electrodes UK Ltd, Celcoat Division, PO Box 111, Highfield Lane, Orgreave, Sheffield S13 9EL, England
Contact: Dr J M Fletcher, Dr C N Richardson
(T: 0742 690771) (Tx: 54293) (Tfx: 0742 693391)

Höganäs Specialty Powder — part of the Kanthal Höganäs Group, a world leading manufacturer of iron and steel powders as well as heating elements, wire and bimetals. The company develops, produces and markets gas atomised powders for thermal coating techniques such as powder welding, flame spraying and plasma welding and for consolidation e.g. hot isostatic pressing and injection moulding. The powders, including self-fluxing, are generally based on nickel, cobalt or iron as well as a wide range of superalloy materials. Customised products and specialist know-how can also be supplied.
Höganäs AB, Business Area Specialty Powder, S-263 83 Höganäs, Sweden
Contact: Michael Bockstiegel, Marketing Manager
(T: 46 42 380 00) (Tx: 72368) (Tfx: 46 42 384 04)

Metco Ltd — part of the international Materials and Surface Technology Group of Perkin-Elmer Corporation, a leading manufacturer and supplier of flame spray equipment and materials. All thermal spray processes are supplied by the company which has been instrumental in major developments in the industry and has approval for supply of equipment and materials to meet military and civil manufacturers' specifications. The company is registered by the British Ministry of Defence as operating quality control procedures in accordance with AWAP 4.
Metco Ltd, Chobham, Surrey GU24 8RD, England
Contact: Eric Brown, Managing Director; W Alan Saywell, Sales Division Executive; R W G Taylor, Company Secretary
(T: 0276 857121) (Tx: 858275 METCO G) (Tfx: 0276 856307)

Mixalloy Ltd — a range of strip metal electrodes for weld cladding using submerged-arc and electroslag techniques including iron-based Feromix, nickel-based Nicromix and cobalt-based Stelmix. These weld cladding alloys are designed to produce wear, corrosion and oxidisation resistant surfaces and a combination of properties can be achieved by varying alloy composition. The company manufactures strip from powder slurry, a process which involves mixing and sintering metal powders. The initial mixing guarantees a uniform distribution of alloying agents and is flexible allowing small quantities to individual specification to be produced in relatively short lead times. Strip is available in widths up to 120mm and 0.5-0.7mm thickness.
Mixalloy Ltd, Antelope Industrial Estate, Rhydymwyn, Mold, Clwyd CH7 5JH, Wales
Contact: Lawton Fage
(T: 0352 741517) (Tx: 617142 MIXLOY G) (Tfx: 0352 741035)

Provacuum Ltd — hardsurfacing of engineering components by thermospraying techniques. Spraying is often followed by fusion in a vacuum furnace to provide a metallurgical bond strength many times higher than mechanical attachment techniques. A range of metal alloys is deposited on a variety of base materials and the company now offers ceramic coatings. In addition vacuum brazing provides similarly strong and uniform bonding between components. Consultancy is offered in choice of unique formulations of coating compounds to meet specific abrasion

and/or corrosion problems. Industries served include oil and gas, chemical, excavation, automotive and engineering.
Provacuum Ltd, 11 Halesfield 18, Telford TF7 4QT, England
Contact: N H Sutherland
(T: 0952 587276) (Tfx:0952 585157)

Soudometal — fluxes and electrodes for arc welding, oriented towards welding and surfacing of special alloys. The Belgium based company, part of the Oerlikon Bührle group, has subsidiaries in Britain and Germany and distributors throughout the world. A complete range of Soudokay flux-cored surfacing wires is offered. Special attention is paid to welding, wear resistant surfacing and cladding techniques for stainless steel, nickel, cobalt, copper and hardfacing alloys. Techniques focus mainly on the covered electrode, submerged-arc and electroslag processes, but flux-cored and solid wires are also available for MIG/MAG and TIG.
Soudometal UK Ltd, 414 Ashton Old Road, Manchester M11 2DT, England
Contact: Alan Burley
(T: 061 2236511) (Tx: 666998 SOUDOM G) (Tfx: 061 2313951)

Index

Abrasive blasting	169
Abrasive wear	19
Accessibility	65
Adhesive wear	21
Aircraft	175
Airframes	176
APS	84
Arc evaporation ion plating	126
Arc wire spraying	83
Austenitic chromium-manganese steel	37
Austenitic irons	37
Austenitic manganese steel	36
Austenitic stainless steels	36
Barrel plating	102, 115
Blank preparation	54, 55
Bond coats	78, 92
Brush or selective plating	103
Buffer layers	52
Bulkwelding	45
Carbon steels	33
Cast irons	49
Cavitation	25
CDS	89
Ceramics	79
Cermets and graded coatings	82
Chemical and petroleum industries	176
Chemical vapour deposition	132
materials and applications	135
reaction types	133
Chromium	106
Chromium boride paste	40
Chromium carbide	79
Coating	
alloys	97
area	62
deposit thickness	33, 53, 55, 91, 106, 107, 110, 111, 113, 141, 142, 143
properties and structure	51, 63
Cobalt alloys	39, 79, 109
Complex irons	37
Compressed gases	163
Consumables	155
Contact fatigue	23
Coolant	148
Copper	108
Copper alloys	39, 109
Corrosion	25, 29
Cracks	67
Cutting tools	146, 147, 150
Cylinders	163
Defects	66, 68, 158, 159, 161
Deposit design	54
Design	
electrodeposition	114
physical and chemical vapour deposition	137
plastics coatings	143
quality assurance	154
weld surfacing	53
Detonation coating	87

Diamond Jet	89
Dilution	50
Distortion control	56
Drilling	147
Earthmoving, agricultural, quarrying and mining	178
Electric shock	168
Electrodeposition	101
coating materials	105
coating selection	113
deposition rates	103, 104
finishing	117
hardness of coatings	106, 108-113
relative costs	114
specifications, inspection and QA	117, 171
substrate materials	116
Electroless or autocatalytic deposition	104
Electroslag	46
Electrostatic spraying	142
Elevated temperatures	25
Enclosed spaces	166
Engines	176
Erosion	25
Explosive weld cladding	68, 69
Ferrous alloys	78
Finishing of welded and sprayed coatings	145, 156
Flame spraying	140
Flames and arcs	164
Fluidised bed coating	141
Flux-cored arc	44
Food processing plant	180
Forging	181
Fretting	24
Friction surfacing	68, 70
adhesion strength	71
applications	73
Fumes and dust	165
Gas ionisation	130
Gas scattering deposition	124
Glass forming tools	182
Gold	111
Grain growth	48
Grinding	148, 150
cylindrical	148
dry and wet	152
parameters	151
safety	170
surface	148
wheel selection	148, 149, 150, 151
Hard anodising	104, 113
Hardness and wear	19
Heat affected zones	47
High chromium irons	37
High speed steels	36
High stress abrasion	19
High velocity combustion spraying	86
Honing	148
Inclusions	68
Inspection	67, 157
Internal combustion engine	179
Interpass temperature control	48
Ion implantation	132
Ion plating	124
Iron	109

Jet Kote	87
Lapping	149
Laser radiation	168
Lead	110
Linishing	152
Low alloy steels	36
Low pressure plasma spraying	85
Low stress abrasion	19
Lubrication	23, 24, 25, 26
Machining	145
allowances	61, 147, 150
Maintenance of equipment	155
Manual metal arc	42
Marine transport	188
Martensitic chromium steels	36
Martensitic irons	37
Masking	93, 103
Materials for weld surfacing	33
Mechanised loading/unloading	65
Mechanised surfacing	63
Metal active gas	44
Metal-ceramic composites	112
Metal inert gas	44
Metal powders	164
Mild and severe wear	22
Milling	147, 152
Mobile plant	179
Nickel	107
Nickel alloys	38, 78
Noise	166
Occupational Exposure Limits	165
Operator training	156
Oxyacetylene	40
Oxyfuel gas powder spraying	83
Oxyfuel gas wire spraying	81
Personal protection	166
Planing and shaping	151
Plasma activated CVD	134
Plasma arc	43, 84
Plastics coatings	139
Platinum	112
Polishing	152
Porosity	67
Porous coatings	150
Post-weld heat treatment	48
Powder welding	42
Power generation	185
Power supplies	46
Precision components	90
Preheat	47
Process plant	177, 179
Protective viewing filters	167
Pulp and paper	182
Quality assurance	153
auditing of quality systems	157
component/substrate requirements	154
consumables	155
finishing, operator training	156
Radial shrinkage	58
Radiation	166
Railways	187

Reactive ion plating	126
Relief cracking	59
Residual stress	47
Resurfacing	33
Road transport	187
Robots	65
Rolling fatigue	23, 24
Rubber	183
Running-in wear	22
Safe working	163, 170
Scuffing	22
Sealing	93
Sewage disposal	181
Silver	111
Small bores, weld surfacing of	66
Solvent degreasing	170
Spark erosion	25, 149
Spray fused coatings	94
Sputter coating	128
Sputter ion plating	127
Steelmaking	185
Stress relief	47
Strip consumables	45
Structure and wear	20
Submerged-arc	45
Substrate materials	48, 96, 116
Surface preparation	91, 97, 116, 120, 138, 143, 169
Surface profile details	90
Surfacing practice	155
Testing and inspection	157, 160
Textiles	186
Thermally sprayed coatings	77
adhesion to substrate	77, 91
evaluation of powders	155
finishing	149
high energy, low energy	77
materials	78
processes	82
surface preparation	91
Three body abrasion	19
Timber	186
Tin	110
Tin alloys	111
Transport	187
Tungsten carbide	40, 79
Tungsten inert gas	43
Turning	146, 150
Vacuum evaporation	122
Vacuum plasma spraying	85
Vapour deposition processes	119
comparison	135
Vat plating	102
Ventilation	165
Vessels with residual flammable material	165
Wear	19
contributory processes	24
mechanisms	19
practical diagnosis	27
resistance testing	26
Welds	
defects	66
patterns	59
shrinkage	47
Welding processes for surfacing	40
Well drilling equipment	177
Work hardening and wear	20